MODELLING WITH DIFFERENTIAL EQUATIONS

MATHEMATICS AND ITS APPLICATIONS

Series Editor: G. M. BELL, Professor of Mathematics,
King's College London, University of London

STATISTICS, OPERATIONAL RESEARCH AND COMPUTATIONAL MATHEMATICS

Editor: B. W. CONOLLY, Emeritus Professor of Mathematics (Operational Research), Queen Mary College, University of London

Mathematics and its applications are now awe-inspiring in their scope, variety and depth. Not only is there rapid growth in pure mathematics and its applications to the traditional fields of the physical sciences, engineering and statistics, but new fields of application are emerging in biology, ecology and social organization. The user of mathematics must assimilate subtle new techniques and also learn to handle the great power of the computer efficiently and economically.

The need for clear, concise and authoritative texts is thus greater than ever and our series endeavours to supply this need. It aims to be comprehensive and yet flexible. Works surveying recent research will introduce new areas and up-to-date mathematical methods. Undergraduate texts on established topics will stimulate student interest by including applications relevant at the present day. The series will also include selected volumes of lecture notes which will enable certain important topics to be presented earlier than would otherwise be possible.

In all these ways it is hoped to render a valuable service to those who learn, teach, develop and use mathematics.

Mathematics and its Applications

Series Editor: G. M. BELL, Professor of Mathematics, King's College London, University of London

Series continued at back of book

MODELLING WITH DIFFERENTIAL EQUATIONS

D. N. BURGHES, B.Sc., Ph.D., F.R.A.S., F.I.M.A.
Cranfield Centre for Teacher Services
Cranfield Institute of Technology

and

M. S. BORRIE, B.A., L.I.M.A.
Pilgrim School
Bedford

ELLIS HORWOOD LIMITED
Publishers · Chichester

Halsted Press: a division of
JOHN WILEY & SONS
New York · Chichester · Brisbane · Toronto

First published in 1981
Reprinted 1982 and 1990 by

ELLIS HORWOOD LIMITED
Market Cross House, Cooper Street, Chichester, West Sussex, PO19 1EB, England

The publisher's colophon is reproduced from James Gillison's drawing of the ancient Market Cross, Chichester.

Distributors:

Australia, New Zealand, South-east Asia:
Jacaranda-Wiley Ltd., Jacaranda Press,
JOHN WILEY & SONS INC.,
G.P.O. Box 859, Brisbane, Queensland 40001, Australia

Canada:
JOHN WILEY & SONS CANADA LIMITED
22 Worcester Road, Rexdale, Ontario, Canada.

Europe, Africa:
JOHN WILEY & SONS LIMITED
Baffins Lane, Chichester, West Sussex, England.

North and South America and the rest of the world:
Halsted Press: a division of
JOHN WILEY & SONS
605 Third Avenue, New York, N.Y. 10016, U.S.A.

British Library Cataloguing in Publication Data
Burghes, David Noel
   Modelling with differential equations.
   (Ellis Horwood series in mathematics and its applications).
   ·1. Differential equations
   I. Title   II. Borrie, Morag S   III. Series
   515.3'5   QA371   80-41936

ISBN 0-85312-286-5 (Ellis Horwood Ltd. – Library Edn.)
ISBN 0-470-27101-9 (Halsted Press – Library Edn.)
ISBN 0-85312-296-2 (Ellis Horwood Ltd. – Student Edn.)
ISBN 0-470-27360-7 (Halsted Press – Paperback Edn.)

Printed in Great Britain by Hartnolls Ltd., Bodmin, Cornwall.

List of Illustrations

Contents

Preface

A few decades ago only physicists and engineers had any use for mathematical analysis, but today the picture has changed dramatically. Mathematics now has important applications in biology, economics, geography, planning, sociology, medicine and psychology. In this book we develop the basic theory of one important branch of mathematics, namely ordinary differential equations, and show its application to many varied disciplines.

The text is written primarily for mathematics teachers and students both at school and in further and higher education. The mathematical level is appropriate for first year undergraduates in a number of disciplines, and for bright scholars in 6th forms. We also hope it will be of benefit to teachers in disciplines other than mathematics, who will gain a greater understanding and appreciation of the power of mathematical analysis.

Most of the material describes other people's work, although frequently adapted and modified to fit the framework of our text. We gratefully acknowledge the help and inspiration we have gained from the books and articles which are referenced at the end of this text. We are particularly grateful to the Open University for permission to use the data described in Section 3.4.

David Burghes
(Cranfield Institute of Technology)
Morag Borrie
(Pilgrim School, Bedford)

Chapter **1**

Introduction

1.1 MATHEMATICAL MODELLING

In this book we will be developing mathematical models which lead to differential equations. We first introduce the concepts involved in mathematical modelling.

The main stages in modelling problems in the real world are illustrated in the diagram below

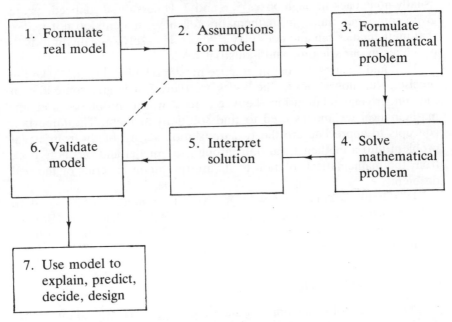

The problem may be to explain some observed data, or make some predictions, or take a decision. To achieve this we translate the real problem into a mathematical one by making a number of simplifying

assumptions. Important variables must be identified, and the relationships between them postulated. The assumptions and relationships constitute the 'mathematical model', and generally lead to a mathematical problem of some sort, which is solved for the relevant variables using appropriate mathematical techniques. The solutions must now be interpreted back in terms of the real problem. Attempts should be made to validate the model, that is to check that the theoretical solution is in good agreement with the observations from the real situation. If there is good correlation, then the model can be used either to give a theoretical explanation for the observed phenomena, to predict further results, or to help in making decisions. On the other hand if the correlation between the theoretical and observed results is not adequate we must return to the assumptions made in the model and decide which need modifying or what additions should be made. The cycle is then traversed once more to see if the new model gives an adequate description of the real problem.

The far left hand column in the diagram represents the real world, the far right hand column the mathematical world and the connections between these two worlds are in the middle column. Most mathematics teaching is solely concerned with box 4, although in applied mathematics there is sometimes a token movement through boxes 1, 2 and 3, but usually none back through boxes 5, 6 and 7. Unless these connections are given some importance in mathematics teaching, students do not realise that mathematics can and does play a vital role in solving many of today's problems in our scientific and industrial world.

It should be stressed that in practice most modelling does not take the precise form shown above. The boxes are there just to give some idea of the underlying relationship between real world problems and the mathematical techniques used to find solutions to them. The important concepts to appreciate are the two translation stages, firstly from a real problem into a mathematical one through the model and secondly back from the mathematical solution to its interpretation in terms of the real problem.

This book is concerned with mathematical models which lead to *ordinary differential equations* at the mathematical problem stage (Box 3). The modelling concepts are illustrated in the next section with an example.

1.2 POPULATION MODELS

The problem of estimating population changes is clearly an important one. The fact that the birth rate in the UK has been in a decline for over a decade has consequences over the next few decades in terms of the country's resources. Fewer maternity beds are required at the first stage, less medical facilities for young children are needed, and the number of new children entering primary school each year decreases. Changes in the way the country's resources are spent need to be planned well ahead, and so good estimates of how the population is changing, not just in overall size, but in age distribution as well, are needed. Similarly towns and cities need to estimate population changes so that suitable facilities and amenities are available.

So our real world problem is to explain the way in which populations change and to formulate a model in order to predict future changes. Our first attempt is based on ideas put forward by the English economist Thomas Malthus in his article 'An Essay on the Principle of Population' published in 1798. If $N = N(t)$ denotes the country's total population at time t, then in a small time interval, say δt, it is assumed that both births and deaths are proportional to the population size and the time interval, i.e.

$$\text{births} = \alpha N \delta t, \text{ deaths} = \beta N \delta t \ (\alpha, \beta \text{ constants}).$$

Thus the increase, say δN, in the total population in the time interval δt is given by

$$\delta N = \alpha N \delta t - \beta N \delta t = (\alpha - \beta) N \delta t = \gamma N \delta t$$

where $\gamma = \alpha - \beta$. Dividing by δt and taking the limit as $\delta t \to 0$ leads to the differential equation

$$\boxed{\frac{dN}{dt} = \gamma N} \tag{1.1}$$

This is a first order differential equation which we can readily solve, since we can rewrite it as

$$\frac{1}{N} \frac{dN}{dt} = \gamma$$

and integrate both sides with respect to t; i.e.

$$\int \frac{1}{N} \frac{dN}{dt} dt = \int \gamma \, dt$$

Thus

$$\int \frac{1}{N} dN = \gamma t + A$$

where A is the constant of integration. Hence

$$\log N = \gamma t + A$$

and if at $t = 0$, $N = N_0$, we have $\log N_0 = A$, and so

$$\boxed{N = N_0 e^{\gamma t}}$$

The predicted behaviour of the population depends very much on the sign of the constant γ. We have exponential growth if $\gamma > 0$, exponential decay if $\gamma < 0$ and no change if $\gamma = 0$. These solutions are illustrated in Fig. 1.1.

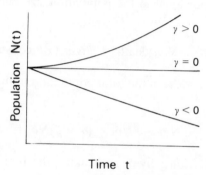

Fig. 1.1—Malthusian Population Model

In order to see if this model is of any value in population predictions, we turn to a specific problem. In Table 1 the USA population statistics

Table 1

Year	USA Population ($\times 10^6$)
1790	3.9
1800	5.3
1810	7.2

are given for 1790, 1800 and 1810. We will use the Malthusian model to predict the future USA population total. Time $t = 0$ corresponds to 1790, so we take $N_0 = 3.9 \times 10^6$. Working in time intervals of 10 years, at $t = 1$ (i.e. 1800), from (1.2)

$$N(1) = 5.3 \times 10^6 = 3.9 \times 10^6 e^{\gamma},$$

so that

$$\gamma = \ln(5.3/3.9) = 0.307$$

Using these values for N_0 and γ, the model predicts $N(2) = 7.3 \times 10^6$ which is reasonably close to the observed value at 1810. Continuing in this way, we calculate the predicted population total every 10 years. These values, together with the actual values, are given in table 2.

Table 2

Year	USA Population ($\times 10^6$)	Predicted Values ($\times 10^6$)
1820	9.6	10.0
1830	12.9	13.7
1840	17.1	18.7
1850	23.2	25.6
1860	31.4	35.0
1870	38.6	47.8
1880	50.2	65.5
1890	62.9	89.6
1900	76.0	122.5
1910	92.0	167.6
1920	106.5	229.3
1930	123.2	313.7

There is reasonable agreement for some time. For example the 1850 figure has a 10% error. But by 1870 the error has increased to 30% and the model is of little use.

We must look back at the assumptions made and consider what factors have been neglected. As it stands the Malthusian Model predicts unlimited growth (for $\gamma > 0$) for all future time. This is most unlikely to occur since there are many varied limitations to growth, such as lack of food resources, overcrowding, insufficient energy supply and other environmental factors.

In 1837 Verhulst proposed a modification which took into account 'crowding' factors. We assume that there is an upper limit, say N_∞, to the population which can be sustained. The population change, dN/dt is now assumed to be proportional to

(i) the current population level, N;
(ii) the fraction of population resource still not utilised, i.e. $(1 - N/N_\infty)$

Thus

$$\frac{dN}{dt} = \gamma N(1 - N/N_\infty) \ . \tag{1.3}$$

This is again a first order differential equation which can be solved in a similar manner to (1.1). We have

$$\int \frac{dN}{N(1 - N/N_\infty)} = \int \gamma \, dt.$$

We can partial fraction the integrand on the left hand side to give

$$\int \left[\frac{1}{N} + \frac{1/N_\infty}{(1 - N/N_\infty)} \right] dN = \gamma t + A$$

where A is the constant of integration. Hence

$$\ln N - \ln(1 - N/N_\infty) = \gamma t + A$$

i.e., $$\ln \frac{N}{(1 - N/N_\infty)} = \gamma t + A.$$

If at $t = 0$, $N = N_0$, we have $A = \ln[N_0/(1 - N_0/N_\infty)]$; and so

$$\frac{N}{(1 - N/N_\infty)} = \frac{N_0 e^{\gamma t}}{(1 - N_0/N_\infty)} \ .$$

Solving for N gives

$$N = N_\infty / \{1 + [(N_\infty/N_0) - 1]e^{-\gamma t}\} \ . \tag{1.4}$$

What does our model now predict? The major change from the Malthusian model is that there is no longer unlimited growth. As $t \to \infty$, we see from (1.4) that

$$N \to N_\infty$$

the maximum sustainable population. How it reaches this level depends on the value of the initial population number. If $N_0 < N_\infty$ there will be exponential type of growth at first, but as N becomes larger, the growth is diminished, and as can be seen from (1.3) as $N \to N_\infty$, $dN/dt \to 0$. A typical

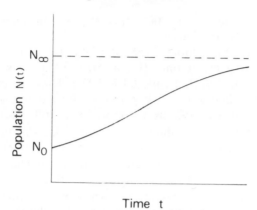

Fig. 1.2—Verhulst Population Model

solution is sketched in Fig. 1.2. On the other hand, if $N_0 > N_x$, there will be initial decay of the population which levels out as N approaches N_x.

We now move onto the validation stage of the modelling process. Again using the USA population data given in Table 1, Verhulst chose the parameters as

$$N_0 = 3.9 \times 10^6, \gamma = 0.3134, N_x = 197 \times 10^6.$$

The predicted values are given in Table 3 together with the actual values up to 1930. There certainly appears to be a remarkable

Table 3

Year	USA Population ($\times 10^6$)	Predicted Values ($\times 10^6$)
1820	9.6	9.7
1830	12.9	13.0
1840	17.1	17.4
1850	23.2	23.0
1860	31.4	30.2
1870	38.6	38.1
1880	50.2	49.9
1890	62.9	62.4
1900	76.0	76.5
1910	92.0	91.6
1920	106.5	107.0
1930	123.2	122.0

correlation, since for over 100 years the Verhulst model accurately predicted the USA population.

So as far as the prediction of the total USA population after 1810, the model looks an excellent one. But after 1930 it does go astray. The limiting value of the Verhulst model is 197×10^6, but already the USA population is past 200×10^6. It is perhaps a little unreasonable to expect the model to predict accurately for more than 100 years. On the other hand, can we have much confidence in using models of this type today to predict the future changes for a country? The laws governing population changes are not as precise as those that govern a mechanical system, and this is why it is difficult to have any real confidence in any population predictions for the future. We do not know what fundamental changes might take place in the next few decades—a nuclear war, a population catastrophe or another industrial revolution?

1.3 A FRAMEWORK FOR MODELLING

In Sec. 1.1 we introduced a framework for the modelling process. Although most modelling will not necessarily follow the exact formulation as illustrated it is instructive to show how the case study in Sec. 1.2 fits into the modelling framework. In fact, in the example we move round the modelling cycle twice. The first circuit can be summarised using the box notation as follows:

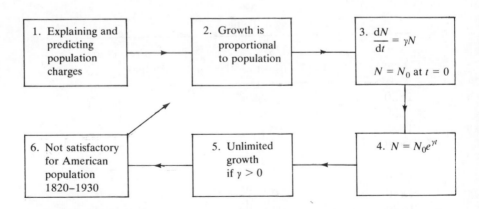

As we were not satisfied with the Malthusian model we went back and looked again at the assumptions made, and so constructed a second model, again moving round the cycle as illustrated below.

Practising applied mathematicians develop a feel for the interplay

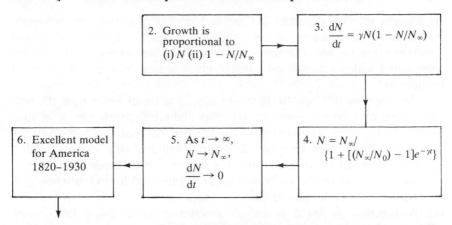

involved between the various stages of modelling and so have no need to work precisely though a scheme as illustrated in the boxes. However, it is a useful framework to build on until one becomes familiar with the modelling process. We will not be using the framework explicitly in the following chapters, but it is the underlying theme for all the examples.

1.4 DIFFERENTIAL EQUATIONS: BASIC CONCEPTS AND IDEAS

By an *ordinary differential equation* we mean a relation which involves one or several derivatives of an unspecified function, say $y = y(x)$, with respect to x; the relation may also involve y itself, and functions of x. For example

$$\text{(i)} \quad \frac{dy}{dx} = ky$$

$$\text{(ii)} \quad \frac{d^2y}{dx^2} + 4y = 0$$

$$\text{(iii)} \quad x^3 \left[\frac{dy}{dx}\right]^2 + 4e^x y = x^2$$

$$\text{(iv)} \quad \frac{d^3y}{dx^3} \cdot \frac{dy}{dx} + y^2 = e^x \frac{dy}{dx}$$

are all ordinary differential equations; whilst, for example

$$\frac{\delta^2 u}{\delta x^2} + \frac{\delta^2 u}{\delta y^2} = 0$$

is an example of a partial differential equation where the unknown variable u is a function of two variables x and y. Partial differential equations are beyond the scope of this book, and so we shall only be concerned with mathematical models which lead to ordinary differential equations.

An ordinary differential equation is said to be of *order n*, if the nth derivative of y with respect to x is the highest derivative of y in that equation. Example (i) and (iii) above are first order, (ii) is second order and (iv) is third order. Chapters 2, 3 and 4 will deal with models leading to first order differential equations, Chapter 5 with models leading to second order differential equations and Chapter 6 with models leading to systems of differential equations.

A function $y = g(x)$ is called a **solution** of a given first order differential equation on some interval, say $a < x < b$ (perhaps infinite), if it is defined and differentiable throughout the interval and is such that the equation becomes an identity when y and dy/dx are replaced by g and dg/dx, respectively. For example, the function $g(x) = e^{\lambda x}$ is a solution of the differential equation

$$\frac{dy}{dx} = \lambda y \tag{1.5}$$

for all x, since $dg/dx = \lambda e^{\lambda x}$ and inserting g and dg/dx in (1.5) reduces it to the identity

$$\lambda e^{\lambda x} = \lambda e^{\lambda x}.$$

Differential equations have in general many solutions. For example

$$y_1 = \sin x, \, y_2 = \sin x + 3, \, y_3 = \sin x - 1 \tag{1.6}$$

are all solutions of

$$\frac{dy}{dx} = \cos x. \tag{1.7}$$

In fact, by integrating both sides of (1.7), we have

$$y = \sin x + c \tag{1.8}$$

where c is an arbitrary constant. So y_1, y_2 and y_3 are all special cases of the general solution, (1.8).

If the solutions of a single differential equation can be represented by a single formula involving arbitrary constants, we call this the **general solution.** If the constants are assigned definite values, then the solutions so obtained are called **particular solutions.** Thus (1.6) are all particular solutions of (1.7). An illustration of the solutions of (1.7) is shown in Fig. 1.3.

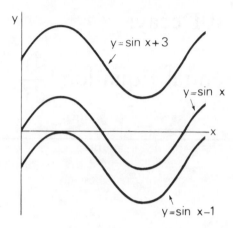

Fig. 1.3—Solutions of $\dfrac{dy}{dx} = \cos x$

It should be noted that all the differential equations that we meet in this book have solutions; but simple equations exist which do not have solutions at all. For example

$$\left[\frac{dy}{dx}\right]^2 = -1$$

has no real solutions. At the beginning of each chapter we outline the theory necessary to follow through the examples though our emphasis is on the uses of differential equations in modelling.

Chapter 2

Growth and Decay:

The Differential Equation '$\dfrac{dy}{dx} = ky$'

2.1 INTRODUCTION

One of the surprises and delights of being an applied mathematician is to recognise that one single mathematical model can represent a vast variety of situations in different disciplines. This is well illustrated by the subject of this chapter, namely models which lead to the differential equation

$$\boxed{\dfrac{dy}{dx} = ky}\,.$$ (2.1)

Once this differential equation has been solved, we have effectively solved numerous problems. For example we will develop mathematical models representing problems in drug absorption, dating of archaeological samples, water cooling and alcohol absorption, all of which reduce to solving a differential equation of the type (2.1). We also set, as case studies for the reader, further applications of this differential equation including population modelling, energy demand forecasting, continuously compounding interest and murder time detection.

We start though by solving the equation, which is a relatively simple one. We can use the technique of separating the variables to give

$$\int \dfrac{dy}{y} = \int k\, dx.$$

Here k is a constant, so we can integrate each side to give

$$\ln y = kx + A,$$

where A is an arbitrary constant. Suppose that at $x = x_0$, $y = y_0$ so that

$$\ln y_0 = kx_0 + A.$$

This determines the value of A and so subtracting the two equations above

$$\ln y - \ln y_0 = k(x - x_0),$$

or

$$\ln(y/y_0) = k(x - x_0).$$

Finally, we can express y as

$$\boxed{y(x) = y_0 e^{k(x-x_0)}}$$ (2.2)

The behaviour of the solution clearly depends on the sign of the constant k. If k is positive we have exponential growth, if k is zero y remains equal to its initial value, and if k is negative we have exponential decay, $y \to 0$ as $x \to \infty$. These solutions are illustrated in Fig. 2.1.

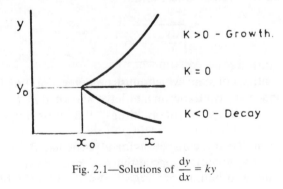

Fig. 2.1—Solutions of $\dfrac{dy}{dx} = ky$

2.2 DRUG ABSORPTION

The study of the way in which a drug loses its concentration in the blood of a patient is fundamental to pharmocology. The 'dose-response' relationship plays a vital role in determining the required dosage level and the interval of time between doses for a particular drug.

Suppose $y = y(t)$ represents the quantity of the drug in the bloodstream at time t. Fig. 2.2 shows some experimental results for penicillin (note that the scale for the concentration is not linear). The

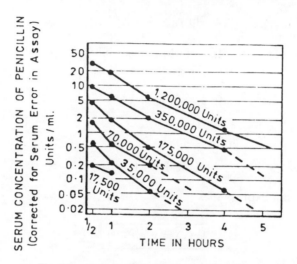

Fig. 2.2—Experimental Results for Penicillin

simplest way to model such behaviour is to assume that the rate of change of the concentration is proportional to the concentration of the drug in the bloodstream. In mathematical terms

$$\boxed{\frac{dy}{dt} = -ky} \qquad (2.3)$$

where k is a positive constant which must be determined experimentally, for the drug under study. Experiments have shown that (2.3) is a good approximation to reality for many drugs, the most important being penicillin.

Having determined the constant k for a particular drug, we now use equation (2.3) as the model. Suppose the patient is given an initial dose y_0, which is assumed to be instantaneously absorbed by the blood at time $t = 0$, resulting in a quantity $y = y_0$ at $t = 0$ in the blood. The actual time for absorption is usually very small compared with the time between doses. From our general solution (2.3), we can immediately write

$$y = y_0 e^{-kt}$$

(where x is replaced by t, x_0 by y_0 and k by $-k$), $-k$ showing that the drug's concentration decays exponentially. After a prescribed time, say T, a second dose, of quantity y_0 is administered. Just before the dose at time $t = T_-$, the amount in the blood is given by

$$y(T_-) = y_0 e^{-kT}$$

Just after the dose, at time $T = T_+$,

$$y(T_+) = y_0 + y_0 e^{-kT}$$

$$= y_0(1 + e^{-kT})$$

What happens to this new quantity in the blood? It decays according to equation (2.3) with initial condition $y = y_0(1 + e^{-kT})$ at $t = T$. Thus for $t \geq T$, from (2.2)

$$y(t) = y_0(1 + e^{-kT})e^{-k(t-T)}.$$

Hence as $t \to 2T$,

$$y(2T_-) = y_0(1 + e^{-kT})e^{kT}.$$

Again giving the patient a dose y_0 at $t = 2T$, results in

$$y(2T_+) = y_0(1 + e^{-kT} + e^{-2kT})$$

and again solving (2.3) with $y = y_0(1 + e^{-kT} + e^{-2kT})$ at $t = 2T$ gives for $t \geq 2T$

$$y(t) = y_0(1 + e^{-kT} + e^{-2kT})e^{-k(t-2T)}.$$

Thus

$$y(3T_-) = y_0(1 + e^{-kT} + e^{-2kT})e^{-kT}$$

and after a dose y_0 at $t = 3T$,

$$y(3T_+) = y_0(1 + e^{-kT} + e^{-2kT} + e^{-3kT}).$$

Continuing in this way,

$$y(nT_+) = y_0(1 + e^{-kT} + e^{-2kT} + \ldots + e^{-nkT}) \qquad (2.4)$$

for $n = 1, 2, \ldots$.

We can now see what is happening to the quantity of the drug as the number of doses increases. For

$$y(nT_+) = y_0 \frac{(l - e^{-(n+1)kT})}{(1 - e^{-kT})} \qquad (2.5)$$

(since (2.4) is a geometrical progression) and as n gets larger, $e^{-(n+1)kT} \to 0$, so that

$$y(nT_+) \to y_0/(1 - e^{-kT}).$$

Since this is independent of n, the model predicts that the quantity of the drug is tending to a saturation level, say y_s, where

$$\boxed{y_s = y_0/(1 - e^{-kT})} \qquad (2.6)$$

This formula plays a key role. It can be used for determining

(i) the required time interval, T, between doses for a given dose y_0 and prescribed final level y_s;
(ii) the dose level y_0 required to obtain a final dose level y_s with a prescribed interval between doses, T.

One disadvantage of this method is the slow build up to the required drug level y_s. Another approach is to start with a large initial dose, say the required final level y_s itself, and then at time T

$$y(T_-) = y_s e^{-kT}.$$

The patient is now given a 2nd dose, say y_d, which brings the level up to y_s again; i.e.

$$y_s = y(T_+) = y_s e^{-kT} + y_d.$$

This gives

$$y_d = y_s(1 - e^{-kT}) = y_0,$$

using (2.6). So we are back to giving a dose y_0 at every time interval T after the initial boost. These methods of accumulation are illustrated in Fig. 2.3.

This second method has the great advantage of reaching the required level immediately, but for many drugs this can have unpleasant side effects on the body. Often in practice a compromise is made between the two methods illustrated above. The patient starts with a double

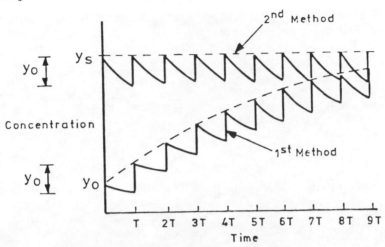

Fig. 2.3—Methods of Drug Accumulation

initial dose, $2y_0$, followed by regular doses y_0. In this way the advantages of each method are used and the disadvantages minimised.

2.3 CARBON DATING

Figure 2.4 shows the famous round table which is fixed on the wall in the great hall of Winchester Castle. The table is 18′ in diameter and has 25 sectors, one for the King and one for each of his knights, the shape being round so that no knight could claim precedence over others. Some experts have thought that this is the actual round table of King Arthur, but there has been recent speculation as to its authenticity. In 1976 the table was taken down from the wall and scientists and archaeologists employed a number of methods to estimate its date of origin. One of the most important methods used was 'Carbon Dating', a technique developed by an American chemist, W. F. Libby, in the late 1940's (for which he received the 1960 Nobel Prize in Chemistry).

Carbon dating is based on the principles of radioactivity discovered and developed at the beginning of the century by Rutherford and others. Certain atoms are inherently unstable so that after some time without any external influence they will undergo transition to an atom of a new element and during transition they emit radiation. From experimental evidence, Rutherford formulated a simple model to describe the way in which radioactive samples decay. If $N = N(t)$ represents the number of atoms in a radioactive sample at time t, then

$$-\frac{dN}{dt}$$

Fig. 2.4—The Round Table in Winchester Castle*

represents the number of disintegrations in unit time. Rutherford showed that the number of disintegrations was proportional to the number of atoms present, so that

$$\frac{\mathrm{d}N}{\mathrm{d}t} = -\lambda N \quad .$$

(2.7)

*Reproduced by permission of Hampshire County Council

Here λ is a positive constant, called the decay constant. This constant will have different values for different substances; the larger the value, the quicker the sample decays. For a particular substance, λ has to be found experimentally. In practice we do not directly measure λ, the decay constant, but the half life, τ. This is defined as the time required for half a given quantity of atoms to decay. To relate λ and τ, we must solve (2.7) with the initial condition $N = N_0$, say at $t = 0$. From (2.2)

$$N(t) = N_0 e^{-\lambda t}, \qquad\qquad (2.8)$$

so that if $N = N_0/2$ at $t = \tau$, we have

$$\frac{N_0}{2} = N_0 e^{-\lambda \tau}.$$

This gives

$$\boxed{\tau = \ln 2/\lambda} \; . \qquad\qquad (2.9)$$

The table below gives some examples of half-lives.

Substance	Half Life
Xenon 133	5 days
Barium 140	13 days
Lead 210	22 years
Strontium 90	25 years
Carbon 14	5568 years
Plutonium	23103 years
Uranium 238	4.5×10^9 years

So for Carbon 16, an isotope of Carbon, the decay constant has value, using (2.9),

$$\lambda = \ln 2/\tau = 1.245 \times 10^{-4} \text{ per year.}$$

We can move onto its application in Carbon-dating. The earth's atmosphere is being continuously bombarded by cosmic rays. This produces neutrons in the atmosphere and these combine with nitrogen to form Carbon-14 (C^{14}). In living plants and animals, the rate of absorption of C^{14} is balanced by the natural rate of decay and an equilibrium state is reached. The basic assumption in carbon dating is that the intensity of bombardment of the earth's surface by cosmic rays has remained constant throughout time. This means that the original rate of disintegrations of C^{14} in a sample of wood is the same as the rate measured today in living wood. When the sample is formed (e.g. the table made) the wood is isolated from its original environment and the C^{14} atoms decay without any further absorption.

Suppose the table was formed at time $t = 0$. Let R_0 denote the original rate of disintegrations, so that

$$R_0 = -\left.\frac{dN}{dt}\right|_{t=0} = \lambda N_0.$$

The present rate of disintegrations, $R(t)$, is given by

$$R(t) = -\frac{dN}{dt} = \lambda N(t) = \lambda e^{-\lambda t} N_0$$

using (2.8). Hence

$$\frac{R_0}{R(t)} = e^{\lambda t},$$

giving

$$t = \frac{1}{\lambda} \ln [R_0/R(t)] \tag{2.10}$$

For living wood, the rate of disintegrations (per minute per gram of sample) is 6.68, so we take this as the value when the table was formed, i.e.

$$R_0 = 6.68.$$

From measurements taken in 1977, the present rate is given by

$$R(t) = 6.08.$$

Using (2.10), the age of the table is given by

$$t = \frac{1}{1.245 \times 10^{-4}} \ln \left[\frac{6.68}{6.08}\right] \approx 700 \text{ years}$$

This gives a date for the table of about 1275, which clearly indicates that the table was not King Arthur's (he lived in the 5th century).

2.4 WATER HEATING AND COOLING

In this section we are concerned with finding optimal policies for the way in which an immersion heater is operated. Is it, for example, cheaper to switch the immersion heater off at night and then on again shortly before it is needed in the morning or allow the thermostat to control the temperature all night? Before looking at the way water cools we first model the way in which water is heated.

Suppose we start with a metal tank full of cold water which is heated by an electric heater fully immersed in the water. To model this situation we will assume the temperature of the water is the same everywhere inside the cylinder. In practice the water will be at a higher temperature near the heater, which often only reaches half way down the cylinder (see Fig. 2.5).

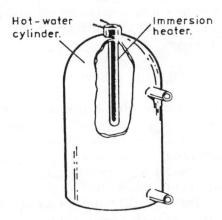

Fig. 2.5—Immersion Heater

We also suppose that heat is produced at a constant rate q for a time T so that the total heat supplied is given by

$$h = qT.$$

Experiments show that the heat required to change the temperature of a mass m of liquid by θ is proportional to both m and θ

i.e. $h \propto m\theta.$

Thus $qT \propto m\theta$ and the constant of proportionality for the liquid is called the specific heat capacity, denoted by c. Hence

$$h = qT = cm\theta \qquad (2.11)$$

Now heat is a form of energy, and the SI unit for energy is the Joule, J. The rate at which energy is produced is called the power, and if energy is supplied at the rate of 1 Joule per second it is said to have power 1 watt, W; i.e.

$$1 \text{ Js}^{-1} = 1 \text{ W}. \qquad (2.12)$$

(For example a 2 kW electric kettle produces 2000J each second). From (2.11) we see that the units for c are J kg^{-1} C^{-1} (where C denotes degrees Centigrade). For water

$$c = 4200 \text{ J kg}^{-1} \text{ C}^{-1} \qquad (2.13)$$

and we can estimate how long it takes an immersion heater, rated at 3 kW, holding 100 kg of water to heat from 15°C to 60°C. Substituting these values in (2.11) gives

$$T = 4200 \times 100 \times 45/3000 = 1\tfrac{3}{4} \text{ hours.}$$

Does this value check with reality? A typical cylinder holds about 100 kg of water, but the time obtained above is probably too long. So what assumptions have we made in the modelling process which were not accurate? We have taken no account of the cylinder's casing, but including it in the model will increase the time taken! Another approximation made was to assume that all the water in the cylinder is heated, but as illustrated in Fig. 2.5, often only about half the cylinder is heated. Since cold water tends to sink, hot to rise, the heater is placed in the top of the cylinder and in effect only heats say a mass $m = 50$ kg of water. This will give a time of heating

$$T = 52\tfrac{1}{2} \text{ mins}$$

which is in better agreement with reality.

To bring the model even nearer to reality we must take into account the casing of the cylinder and the surroundings. Now heat supplied by heater = heat given to water + heat given to casing + heat given to surroundings.

If the casing is assumed to be at the temperature of the water with which it is in contact, and m_c is the mass of that part of the casing surrounding the heated water, then

$$h = c_w m_w \theta + c_c m_c \theta \qquad (2.14)$$

where suffix w refers to the water and c_c is the specific heat capacity of the casing, and we have neglected heat lost to the surroundings. For example if $c_c = 400$ J kg^{-1} C^{-1}, and $m_c = 5$ kg, then the time taken to raise the water temperature from 15°C to 60°C is now given by

$$T = (4200 \times 50 + 400 \times 5) \times 45/3000 = 53\tfrac{1}{2} \text{ mins.}$$

So consideration of heat lost to the casing appears to be relatively insignificant.

We are now in a position to consider the most economical way to operate an immersion heater. To analyse the problem we first consider what happens when the heater is switched off and the water cools. To model the cooling of water we suggest you try the following experiment.

Experiment

Take a cup of very hot water and place a thermometer in the liquid. Read the temperature initially and thereafter every 20 minutes for two hours. Plot the data on a graph, drawing a smooth curve to connect the data points.

Measure the slope of the curve at every 20 minute point and plot this against the temperature.

The final graph should give an approximate straight line and illustrates Newton's law of cooling, which states the temperature difference, θ, between the hot body and its surroundings decreases at a rate proportional to the temperature difference. In mathematical terms

$$\boxed{\frac{d\theta}{dt} = -k\theta} \qquad (2.15)$$

where k is a positive constant. Returning to the situation of the water cooling inside the immersion heater, we must find the appropriate form for the constant k.

If q now stands for the rate of loss of heat, an alternative way of expressing the law of cooling is to assume that q is proportional to the surface area, a, and the temperature difference, θ. Thus

$$q = ja\theta$$

where the constant j is called the heat transfer coefficient, which depends on the nature of the surface of the hot body and the surrounding environment. Now if h is the total heat originally supplied, then

$$\frac{dh}{dt} = -ja\theta.$$

But from (2.14), $h = (c_w m_w + c_c m_c)\theta$ giving

$$\frac{d\theta}{dt} = \left[\frac{-ja}{c_w m_w + c_c m_c}\right]\theta \qquad (2.16)$$

so the parameter k in (2.15) has value $ja/(c_w m_w + c_c m_c)$.

The solution of (2.16) is illustrated in Fig. 2.6. At $t = 0$, it is assumed that $\theta = \theta_0$ giving

$$\boxed{\theta(t) = \theta_0 e^{-kt}} \qquad (2.17)$$

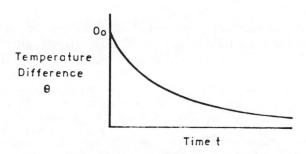

Fig. 2.6—Temperature Difference Against Time

Knowing that θ decays exponentially, we can see that the slope of the curve is much greater when θ is large than when it is small. Hence keeping the water hot means that the water is always cooling at its greatest rate, the loss of heat being rectified by regular boosts from the heater. On the other hand, allowing the water to cool until just before hot water is needed again means that the cooling takes place progressively more slowly, shown by the slope of the curve becoming less steep. The loss of heat over the period of time must therefore be less, and so less heat in total needs to be supplied by the immersion heater.

So by graphical arguments it is clear that it is cheaper to switch off the heater until the water is needed. But how much cheaper?

Method 1: Suppose the immersion heater as described earlier has been maintaining the water in the cylinder at 45°C above the surrounding temperature. If it is switched off at 11 p.m. at night, then from (2.17) the temperature difference at 7.00 a.m. in the morning is given by

$$45e^{-28800k}$$

If $a = 1$ m^2, $j = 10$ W m^{-1} C^{-1}, then using values as above, we have

$$k = 10 \times 1/(50 \times 4200 + 5 \times 400) = \frac{1}{21200}.$$

This gives a temperature difference of approximately 12°C, showing that the water has dropped by $45 - 12 = 33$°C.

Using (2.14) and (2.11), the time taken to raise the temperature by 33°C is

$$T = \frac{(c_w m_w + c_c m_c)\theta}{q} = \frac{(4200 \times 50 + 400 \times 5)33}{3000} \text{ s}$$

which gives $T \simeq 40$ min. A 3 kW heater uses $3 \times 2/3 = 2$ units of electricity when switched on for 40 minutes (one unit of electricity is consumed by a device rated at 1 kW in one hour). So this method used approximately 2 units of electricity.

Method 2: This time we allow the thermostat to control the temperature throughout the night. Suppose that the thermostat switches on at 42°C above the environment and off again at 45°C. The time taken to raise the temperature by 3°C is given by

$$T = \frac{(c_w m_w + c_c m_c)\theta}{q} = \frac{(4200 \times 50 + 400 \times 5)3}{3000} \text{ s}$$

i.e. $T \simeq 3\frac{1}{2}$ mins. The time taken to cool from 45°C to 42°C is given from (2.17) by

$$t = \frac{1}{k} \ln \frac{\theta_0}{\theta(t)} = 21200 \ln \left[\frac{45}{42}\right]$$

i.e. $t = 24\frac{1}{2}$ mins. So the system is continually going through a cooling–heating cycle every 28 mins. This is illustrated in Fig. 2.7.

In the 8 hrs 40 mins when the system regulated by method 1 is cooling and reheating once, the system regulated by method 2 has gone

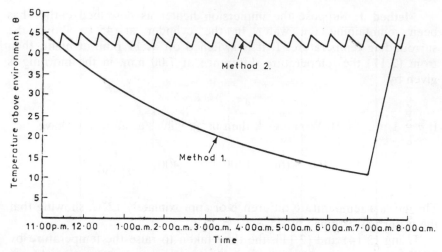

Fig. 2.7—Methods of Operating an Immersion Heater

through about 19 cycles, and so the heater has been on for $(19 \times 3\frac{1}{2})$ minutes = $66\frac{1}{2}$ minutes. Thus the 3 kW heater uses $3 \times 66.5/60$ units = 3.325 units of electricity.

So the first method saves at least one unit of electricity each night, at a cost of about 4p a unit. Perhaps not a very significant saving, but worth £14 a year!

2.5 ALCOHOL ABSORPTION: ACCIDENT RISK

Figure 2.8 below illustrates the relationship between the relative risk of having a car accident, R, and the blood alcohol level, b. The points indicated have been found from extensive research work and the curve drawn is based on the modelling assumption that

$$\boxed{\frac{\mathrm{d}R}{\mathrm{d}b} = kR} \qquad (2.18)$$

where k is a positive constant. If at $b = 0$ (no alcohol at all) the risk of an accident is 1%, i.e. $R_0 = 1$, then from (2.2) the solution of (2.18) is given by

$$R(b) = e^{kb}. \qquad (2.19)$$

So the model predicts an exponential increase in the accident risk with increasing blood alcohol level.

l

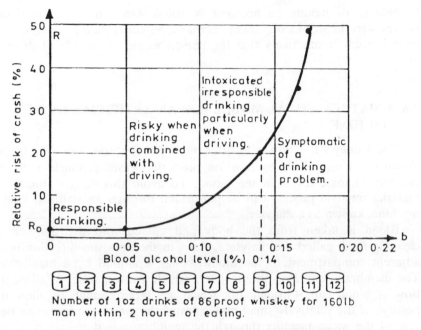

Fig. 2.8—Blood Alcohol Level and Accident Risk

To validate the model we must use the available data to estimate the constant k. For example using the data point $R = 20\%$ at $b = 0.14\%$ gives

$$k = \frac{1}{0.14} \ln 20 = 21.4.$$

So the solution of our model is given by

$$\boxed{R(b) = e^{21.4b}}$$ (2.20)

This is the curve illustrated in Fig. 2.8 and it is clearly a good fit to the data points.

One limitation to the model is that as b increases, the risk level increases far beyond 100%. For example if $R = 100$ which means a crash is certain to occur, the corresponding value of the blood alcohol level is

$$b = \frac{1}{21.4} \ln 100 = 0.22.$$

So according to the model, when the blood alcohol level is 0.22%, the probability of having an accident is 100%. On reflection, since this occurs after 12 drinks of whisky, perhaps the conclusion is a reasonable one! It seems more likely that the person would be unable to drive at all.

2.6 A MATHEMATICAL MODEL FOR AN ARTIFICIAL KIDNEY MACHINE

The kidney's function is to filter out waste material from the blood. When the kidney fails to function properly, waste products build up possibly to toxic levels in the blood. To avoid this danger, the waste material removal process can be performed through an artificial kidney machine, known as a dialyser.

Blood is taken from the body and passed into the dialyser. A cleaning fluid, called the dialysate, flows in the opposite direction in an adjacent compartment to the blood, being separated by a membrane. The membrane contains minute pores which are too small to allow the flow of blood cells through, but which are large enough to allow the passage of the relatively small molecules of the waste product. The flow rate of the waste product through the membrane is determined by the differences in concentration on either side, the flow being from high to low concentration. The situation is illustrated in Fig. 2.9.

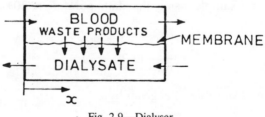

· Fig. 2.9—Dialyser

The important quantity is the removal rate which will depend on the flow rates of the blood and dialysate through the dialyser, the size of the dialyser and the permeability of the membrane. We will take the last two factors as fixed, and concentrate on finding the dependence of the removal rate on the flow rates. Let x denote the distance along the dialyser. We consider what happens in a small section of the dialyser from x to $x + \delta x$, as shown in Fig. 2.10.

The key variables are the concentrations of the waste product in the blood and dialysate. Denoting these by u and v respectively, we assume that these quantities are functions of x, the distance along the dialyser, i.e. $u = u(x)$, $v = v(x)$.

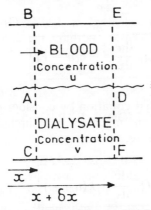

Fig. 2.10—Cross-Section of Dialyser

The experimental law governing the amount of material passing through the membrane is **Fick's law** which states that

'the amount of material passing through the membrane is proportional to the difference in concentration.'

The difference in concentration across BC is $u(x) - v(x)$, and so the mass transfer through a section of membrane of unit width and length δx from blood to dialysate in unit time is approximately given by

$$k[u(x) - v(x)]\delta x.$$

The proportionality constant is assumed independent of x. To derive the governing equations of the model we must consider the mass changes in the element $BEFC$ in unit time. Now in unit time,

| mass flow across AB into element | = | mass passing through membrane AD | + | mass flow across ED out of element |

and putting this into mathematical terms,

$$Q_B u(x) = k[u(x) - v(x)]\delta x + Q_B u(x + \delta x) \qquad (2.21)$$

where Q_B is the flow rate of blood through the machine. Thus

$$Q_B \left[\frac{u(x + \delta x) - u(x)}{\delta x} \right] = -k[u(x) - v(x)]$$

and letting $\delta x \to 0$ gives

$$Q_B \frac{du}{dx} = -k(u - v) \ . \tag{2.22}$$

We obtained the differential equation by considering a small element of the blood flowing through the machine. In a similar way if we consider a small element of the dialysate, we obtain

$$-Q_D \frac{dv}{dx} = k(u - v) \ . \tag{2.23}$$

Equations (2.22) and (2.23) are the governing coupled differential equations for the model. We can solve them be adding (2.22) and (2.23) to give

$$\frac{du}{dx} - \frac{dv}{dx} = -\frac{k}{Q_B}(u - v) + \frac{k}{Q_D}(u - v),$$

so that if $z = u - v$,

$$\frac{dz}{dx} = -\alpha z \tag{2.24}$$

where $\alpha = k/Q_B - k/Q_D$. We are back again to equation (2.1) which in this case has solution

$$z = Ae^{-\alpha x}, \tag{2.25}$$

A being an arbitrary constant. Returning to (2.22) we see that

$$\frac{du}{dx} = -\frac{k}{Q_B} z = -\frac{k}{Q_B} Ae^{-\alpha x}$$

and integrating gives

$$u = B + \frac{kA}{\alpha Q_B} e^{-\alpha x} \tag{2.26}$$

where B is an arbitrary constant. We can now obtain v from (2.25) using (2.26), since

$$u - v = Ae^{-\alpha x}$$

i.e.

$$v = B + \frac{kA}{\alpha Q_D} e^{-\alpha x}. \qquad (2.27)$$

To complete the solution we must specify the boundary conditions. Suppose that the blood has initial concentration u_0 on entry and that the dialysate has almost zero concentration on entry; i.e.

$$u = u_0 \text{ at } x = 0$$

$$v = 0 \text{ at } x = L$$

where L is the length of the dialyser. Applying these conditions determines A and B and after some algebra the final solutions are

$$u = u_0 \left[\frac{(e^{-\alpha L}/Q_D) - (e^{-\alpha x}/Q_B)}{(e^{-\alpha L}/Q_D) - (1/Q_B)} \right]$$

$$v = \frac{u_0}{Q_D} \left[\frac{e^{-\alpha L} - e^{-\alpha x}}{(e^{-\alpha L}/Q_D) - (1/Q_D)} \right]. \qquad (2.28)$$

We now interpret our solution in terms of the important factor in the dialyser. Clearly the most important factor is the amount of waste material removed (in unit time). This quantity is predicted as

$$\int_0^L k[u(x) - v(x)] \, dx$$

$$= -Q_B \int_0^L \frac{du}{dx} \, dx, \text{ using (2.22)}$$

$$= -Q_B \int_{u_0}^{u(L)} du$$

$$= -Q_B[u_0 - u(L)].$$

Dialyser designers work in terms of the 'Clearance', Cl, which is defined by

$$Cl = \frac{Q_B}{u_0}[u_0 - u(L)].$$

Using (2.28), we obtain eventually

$$Cl = Q_B\left[\frac{1 - e^{-\alpha L}}{1 - (Q_B/Q_D)e^{-\alpha L}}\right],$$

where

$$\alpha L = \frac{kL}{Q_B}(1 - Q_B/Q_D).$$

The key parameters for this model are

    (i) Q_B/Q_D, the flow rate ratio

    (ii) kL/Q_B.

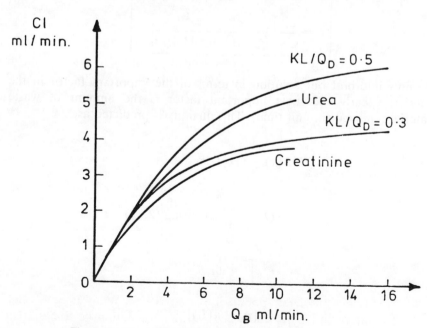

Fig. 2.11—Experimental/Theoretical Curves for the Clearance

In typical operating conditions, Q_B varies from 100 to 300 ml/min and Q_D from 200 to 600 ml/min, whilst the ratio kL/Q_B is in the region 1 to 3. Before being able to use this model to help design dialysers we must test it against experimental data. If, for example, k, L and Q_D are kept constant, but Q_B is varied, we obtain the curve shown in Fig. 2.11. Also shown are two experimental curves and there is reasonable agreement between the theoretical and experimental curves. So our model appears to be a reasonable one but a more realistic model is required. The next stage in the modelling process would be to improve the model by taking into account such factors as variations of k with x, the depths of the blood and dialyser channels and pressure differences across the membrane.

EXERCISES

1. **Carbon Dating**

   Charcoal from the famous Lascaux Cave in France gave an average count in 1950 of 0.97 disintegrations (per minute per gram). Living wood gave 6.68 disintegrations. Estimate the date of formation of the charcoal and give a date to the remarkable paintings in the cave.

2. **Saving Electricity**

   Estimate the number of units of electricity required during the night using Methods 1 and 2 (as described in Section 2.4) in order to keep the water temperature at $40°C$ above the environment. Determine the money saved over a year using this temperature rather than $45°C$. [Use all required data from Section 2.4.]

3. **Population Growth**

   (i) The population of USSR was 209 million in 1959, and it was estimated to be growing exponentially at a rate of 1% per year. This means

   $$\frac{dP}{dt} = (0.01)P.$$

   Find the predicted population after 1959. What is the value predicted for 1980, and when will the population be double that of 1959?

(ii) The population of New Zealand is given in the table below.

Year	Population
1921	1.218×10^6
1926	1.344×10^6

Modelling the change in population by the differential equation

$$\frac{dP}{dt} = kP,$$

use the data above to estimate the value of k. Predict the population in 1936, 1945, 1953 and 1977, and compare with the actual data given below.

Year	Population
1935	1.491×10^6
1945	1.648×10^6
1953	1.923×10^6
1977	3.140×10^6

What conclusions regarding the suitability of your model can you reach?

4. **Continuous Interest**

A building society advertises that interest will be compounded continuously at a rate of 10% per year. This means that if P is the balance in an account at time t,

$$\frac{dP}{dt} = (0.1)P.$$

If P_0 is invested on the first day of a year, find how much is in the account at the end of the first year. What is the 'effective' yearly interest rate? How long will it take an investment of £100 to double itself?

5. **Estimation of Time of Murder**
   The body of a murder victim was discovered at 11.00 p.m. one evening. The police doctor on call arrived at 11.30 p.m. and immediately took the temperature of the body, which was 94.6°F. He again took the temperature after one hour when it was 93.4°F, and he noted that the temperature of the room was a constant 70°F. Use the law of cooling (as described in Section 2.4) to estimate the time of death.

6. **Weight Loss**
   In a fasting experiment, the weight of a volunteer decreased from 180 lbs to 155 lbs in 30 days. It was noted that the weight loss per day was approximately proportional to the weight of the volunteer. Determine a differential equation to describe this behaviour and estimate how long it will take the volunteer to reach 130 lbs.

7. **Energy Demand**
   (i) Past data on world oil production is given below

Year	Production (millions of metric tons)
1960	1091
1962	1259
1964	1458
1966	1693

   Assuming that the rate of growth of production is proportional to the production, estimate the production in 1968, 1970 and 1972. Compare your predictions with the actual values given below.

Year	Production
1968	1986
1970	2340
1972	2550

(ii) USA petrol consumption (in 1000 barrels per day) is given below.

Year	Consumption
1946	2015
1950	2724
1955	3655
1960	4130

Formulate a differential equation model to describe the growth in petrol consumption and use it to estimate the 1965 and 1970 values. These values are actually 4853 and 6083 respectively. If you think your model is a reasonable one, estimate the 1972 value.

[The actual 1972 value is 6668]

Chapter **3**

Variables Separable Differential Equations

3.1 INTRODUCTION

In this chapter we consider a generalisation of the differential equation introduced in Chapter Two. We develop models which lead to a differential equation of the form

$$\boxed{\frac{dy}{dx} = f(y)\, g(x)}.$$ (3.1)

Note that in the special case $f(y) = Ky, g(x) = 1$, we are back to $dy/dx = Ky$, i.e. equation (2.1).

Now the solution of differential equation (3.1) is found by separating the variables to give

$$\int \frac{dy}{f(y)} = \int g(x)dx.$$ (3.2)

Provided we can integrate both sides we can solve such equations, but the form of the solution will depend on the actual functions f and g.

Example: If $f(y) = 1/y$ $(y \neq 0)$, $g(x) = -x$, then

$$\int y\, dy = -\int x\, dx.$$

Integrating, $\frac{1}{2}y^2 = -\frac{1}{2}x^2 + A$, where A is the constant of integration

i.e. $x^2 + y^2 = 2A$.

If also $y = 1$ when $x = 1$, then $2A = 2$ and $x^2 + y^2 = 2$ and the solution curve is a circle, centre the origin, and radius $\sqrt{2}$, as illustrated in Fig. 3.1.

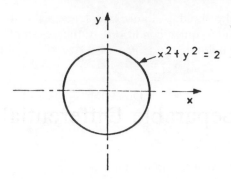

Fig. 3.1—Solution Curve for $\dfrac{dy}{dx} = -\dfrac{x}{y}$, $y(1) = 1$

3.2 REACTION TO STIMULUS

Our sensory organs are responsive to a wide range of stimuli. For example the ear can detect sound ranging in intensity from a pressure of 0.0002 units of force (dynes) per square centimetre to a pressure of 2000 dynes per cm^2. This means the ear can detect sounds from as low as a ticking watch twenty feet away to noises as loud as a jet aircraft.

The first mathematical model to describe the response, R, to a stimulus S, was due to the German physiologist Gustav Fechner (1801–1887). The model can be written in differential form

$$\boxed{\dfrac{dR}{dS} = \dfrac{k}{S}} \tag{3.3}$$

where k is some positive constant. This equation implies that the increment to the reaction for equal increments in stimulus decreases as the magnitude of the stimulus increases. For example a small noise when lying awake at night, when the background noise is low, is quite significant whereas the same noise would go unobserved during the daytime when the background noise level is much higher.

Equation (3.3) is a variable separable differential equation and we can write its solution as

$$\int dR = \int \dfrac{k}{S}\, dS$$

i.e. $$R = k \ln S + A, \tag{3.4}$$

where A is the constant of integration. Now let S_0 be the lowest level of the stimulus which can be consistently detected. We call this the threshold value or detection threshold. As an example we can take the tick of a watch at 20 feet under quiet conditions. Other examples of detection thresholds are given below.

Stimulus	Detection Threshold
Light	The flame of a candle 30 miles away on a dark night
Taste	Water diluted with sugar in the ratio of 1 teaspoon to two gallons
Smell	One drop of perfume diffused into the volume of three average size rooms
Touch	The wing of a bee dropped on your cheek at a distance of 1 centimetre (about 3/8 of an inch)

Hence we take the reaction to the detection threshold as zero i.e. $R(S_0) = 0$. Using this in (3.4) gives $A = -k \ln S_0$ and finally

$$R = k \ln(S/S_0)$$. (3.5)

A typical solution is illustrated in Fig. 3.2. Clearly the values of the parameters k and S_0 depend on both the type of stimulus and the individual.

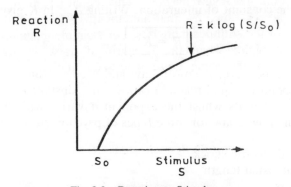

Fig. 3.2—Reaction to Stimulus

As expected the solution tells us that the increment in the reaction becomes smaller as one receives stronger stimulus. For example if you are in an average size room with just one 50 watt bulb in one lamp, an increase of 50 watts to 100 watts would produce a dramatic improvement in the quality of the light provided; but the same increase of 50 watts to 150 watts would make only a small improvement. The Weber-Fechner law essentially tells us that sensations only increase in an arithmetic form in response to logarithmic changes in the stimulation. This explains the remarkable capacity of the ear to respond efficiently to such a wide range of stimuli.

For some time this model was regarded as a basis for psycho-physics; but in the early 1950's new methods were devised for measuring sensory perception. These were largely conceived by S. Stevens (1906–1973) and he first showed that the Weber-Fechner Law was inadequate. He formulated his own law, which is based on the differential equation

$$\frac{dR}{dS} = n\frac{R}{S} \qquad\qquad (3.6)$$

where n is a positive constant. Note that this is the same as (3.3) except for the addition of an R term on the right hand side. So we are now saying that the rate of change of reaction to stimulus varies inversely with stimulus (as before) but also linearly with the reaction. We can readily integrate (3.6) to give

$$\int \frac{1}{R}\, dR = \int \frac{n}{S}\, dS$$

i.e. $\ln R = n \ln S + A$

where A is the constant of integration. Writing $A = \ln K$ gives

$$\boxed{R = KS^n} \; . \qquad\qquad (3.7)$$

This is known as Stevens' Power Law and predicts that equal stimulus ratios correspond to equal reaction ratios. The constant K is determined by the choice of units whilst the exponent n varies with the source of sensation. The power law for three types of psychological sensations

  (i) perceived electric shock
 (ii) apparent visual length
(iii) brightness

is illustrated in Fig. 3.3. Stevens' data appears to demonstrate that the power law (3.7) accurately describes the relationship between stimulation and sensation.

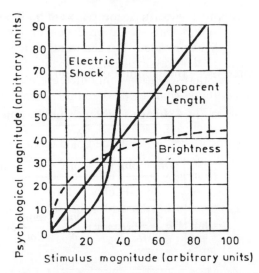

Fig. 3.3—Experimental Results for Reaction to Stimulus

3.3 ROCKET FLIGHT

The launching of yet another satellite no longer provides the news headlines (unless it is a failure); it is just a routine part of our life now, very unlike the 4 October 1957 when the first earth satellite, Sputnik 1, was launched by Russia. Since then rocket technology has progressed at an enormous pace. The Saturn rockets used for the Apollo moonshots were essentially three stage rockets together with the satellite and its lunar module.

In this section we will first look at the main characteristics of rocket flight, and then consider the problem of launching a satellite into a typical orbit about the earth. Rockets are continuously losing mass, in fact not just losing it but it is being propelled away at significant speeds; we must be careful in applying Newton's second law (force is proportional to rate of change of linear momentum) and consider a rocket of mass m, moving with speed v which in a small time δt loses a small mass, say δm_p, which leaves the rocket with speed u in the opposite direction to v. The situation is illustrated in Fig. 3.4. The resulting speed of the rocket is $v + \delta v$.

Fig. 3.4—Rocket System in Time δt

We can now apply Newton's second law of motion to the whole system, giving

$$\text{force} = \frac{d}{dt}(\text{momentum of system})$$

$$= \lim_{\delta t \to 0}\left[\frac{\text{change in momentum in time } \delta t}{\delta t}\right]$$

$$= \lim_{\delta t \to 0}\left[\frac{[(m - \delta m_p)(v + \delta v) + \delta m_p(-u)] - mv}{\delta t}\right]$$

$$= \lim_{\delta t \to 0}\left[m\frac{\delta v}{\delta t} - (u + v)\frac{\delta m_p}{\delta t} - \frac{\delta m_p}{\delta t}\delta v\right]$$

i.e.

$$\boxed{F = m\frac{dv}{dt} - (u + v)\frac{dm_p}{dt}} \tag{3.8}$$

where F is the force acting on the system in the v direction. We usually write $c = u + v$, so that c, called the relative exhaust speed, is the speed of the propellant (burnt gases) relative to the rocket. Also the term dm_p/dt is the positive rate of change of the propellant mass, and so

$$\frac{dm_p}{dt} = -\frac{dm}{dt}.$$

Thus for motion in one direction, (3.8) can be written as

$$F = m\frac{dv}{dt} + c\frac{dm}{dt}. \tag{3.9}$$

But what are the main forces acting on the rocket as it leaves the earth's surface? The gravitational pull is the dominant one, and in the initial motion the drag due to the earth's atmosphere will be important although this diminishes fairly rapidly with height above the surface. On the other hand we are just trying to get an overall picture of rocket flight, so initially we neglect all external force; i.e. $F = 0$ in (3.9). Hence our model is governed by the differential equation

$$\frac{dv}{dt} = \frac{c}{m}\frac{dm}{dt}$$

which we can express as

$$\boxed{\frac{dv}{dm} = -\frac{c}{m}} \tag{3.10}$$

regarding v as a function m. This is a variable separable equation, in fact of the same form as (3.3), the Weber-Fechner differential equation.

The solution of the differential equation is

$$v = -\int \frac{c}{m} \, dm + A$$

i.e. $$v = -c \ln m + A,$$

A being the constant of integration. Initially suppose $v = 0$ and $m = m_0 + P$, where P is the payload (e.g. satellite) mass and m_0 the initial rocket mass (excluding the satellite). This mass is divided into two parts, the initial fuel mass, say ϵm_0 $(0 < \epsilon < 1)$, and the casing and instruments of mass $(1 - \epsilon)m_0$. We call $(1 - \epsilon)$ the *structural factor* of the rocket. The constant k is determined from $0 = c \ln(m_0 + P) + k$, so that

$$v = -c \ln[m/(m_0 + P)].$$

When all the fuel has been burnt, $m = (1 - \epsilon)m_0 + P$, so that

$$\boxed{v = -c \ln[1 - \epsilon/(1 + \beta)]} \tag{3.11}$$

where $\beta = P/m_0$.

This final speed is very much an upper bound to the possible speed that can be given to the satellite. The forces that have been neglected all reduce this final speed. Nevertheless it is instructive to evaluate (3.11) for a typical rocket. The predicted final speed (3.11) depends on the three parameters

$$c, \epsilon \text{ and } \beta$$

The relative exhaust speed c, for both liquid and solid fuel rockets, has a typical value of about 3.0 km s^{-1}. The value of ϵ depends on the materials used but is typically about 0.8. The value of β, the ratio of the payload mass to the rocket mass, will be small, say about 1/100. Using these values, (3.11) gives

$$v_1 \simeq 4.7 \text{ km s}^{-1}.$$

So this is an upper estimate to the typical final speed which a single stage rocket can give to its payload.

The first problem considered by scientists planning to launch satellites was to consider whether this could be achieved using a single stage rocket. To find out we must determine the speed that a satellite needs to have in order to stay in a circular orbit of height h above the earth's surface, radius R_e, as illustrated in Fig. 3.5. The gravitational pull towards the

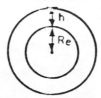

Fig. 3.5—Satellite in Circular Orbit About Earth

centre of the earth is given by Newton's inverse square law of attraction as

$$\gamma m M_e / a^2$$

where γ is the universal gravitational constant, m the mass of the satellite, M_e the earth's mass and $a = h + R_e$, the radius of the orbit. This force is balanced by the centrifugal force mv^2/a, v being the satellite's speed. Thus

$$\gamma m M_e / a^2 = mv^2 / a$$

giving

$$v = [\gamma M_e / (h + R_e)]^{1/2}. \qquad (3.12)$$

Now for a typical orbit of height 100 km above the earth's surface,

$$v \simeq 7.8 \text{ km s}^{-1}.$$

So in order to launch such a satellite, the rocket must be able to reach speeds of this magnitude.

But this is far in excess of the value obtained from a single stage rocket. So we next turn to multi-stage rockets in which after each stage has been burnt, the casing of that stage is discarded. In this way the unwanted casing mass is not carried throughout the motion and so we would clearly expect a better final speed. How much better though? Let's consider a two stage rocket, with stages of initial mass m_1 and m_2 respectively. For simplicity we assume equal relative exhaust speeds, c, and equal structural factors. The two stage rocket is illustrated in Fig. 3.6.

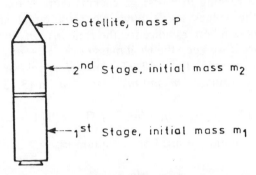

Fig. 3.6—Two Stage Rocket

For the first stage burn the payload carried is the second stage and the satellite, i.e. $m_2 + P$. Thus from (3.11) with $\beta = (m_2 + P)/m_1$, the speed obtained by the first stage is given by

$$- c \ln[1 - \epsilon m_1/(m_1 + m_2 + P)].$$

The second stage is just a single stage rocket with initial mass m_2 and payload P, and again using (3.11) this stage provides a further increment in speed

$$- c \ln[1 - \epsilon m_2/(m_2 + P)].$$

Thus the final speed given to the satellite is given by

$$\boxed{V_2 = - c \ln[1 - \epsilon m_2/(m_1 + m_2 + P)] - c \ln[1 - \epsilon m_2/(m_2 + P)]}.$$
$$(3.13)$$

To compare with the single stage rocket considered above we take, for example,

$$\epsilon = 0.8, c = 3.0 \text{ km s}^{-1}, m_1 = m_2 = 50P$$

so that $$P/(m_1 + m_2) = 1/100;$$

and with these values $$v_2 \simeq 6.2 \text{ km s}^{-1}.$$

So although we have obtained a considerable improvement over the single stage rocket ($v_1 \simeq 4.7$ km s^{-1}) it is still not sufficient to launch the satellite.

But before rushing to a 3 stage rocket, there is one more action we can try. In the values used above we took equal stage masses, i.e. $m_1 = m_2$. Is there a better choice for the ratio $m_1 : m_2$ than 1 : 1? Will the final speed vary if we keep the total mass $m_1 + m_2 = m_0$, say, constant but allow the ratio $m_1 : m_2$ to vary? The answer is yes, and to see how to best choose the ratio $m_1 : m_2$ put $m_1 = m_0 - m_2$ in (3.13). Then

$$v = -c \ln[1 - \epsilon(m_0 - m_2)/(m_0 + P)] - c \ln[1 - \epsilon m_2/(m_2 + P)]$$

and so v_2 is a function of m_2. For maximum v_2,

$$\frac{dv_2}{dm_2} = 0,$$

which gives

$$\frac{\epsilon/(m_0 + P)}{[1 - \epsilon(m_0 - m_2)/(m_0 + P)]} = \frac{\epsilon P/(m_2 + P)^2}{[1 - \epsilon m_2/(m_2 + P)]} .$$

Cross multiplying and simplifying gives

$$(1 - \epsilon)(m_2^2 + 2Pm_2 - Pm_0) = 0.$$

Since $\epsilon \neq 1$, we must have

$$m_2^2 + 2Pm_2 - Pm_0 = 0$$

and solving

$$m_2 = -P + (P^2 + Pm_0)^{1/2}.$$

Thus with $\beta = P/m_0$,

$$\frac{m_2}{m_0} = -\beta + (\beta + \beta^2)^{1/2}$$

and since $m_1 = m_0 - m_2$,

$$\frac{m_1}{m_0} = 1 + \beta - (\beta + \beta^2)^{1/2} = (1 + \beta)^{1/2}[(1 + \beta)^{1/2} - \beta^{1/2}].$$

Thus for maximum final speed, we must choose the ratio m_1/m_2 as

$$\frac{m_1}{m_2} = \frac{(1 + \beta)^{1/2}}{\beta^{1/2}}$$

i.e.

$$\boxed{m_1/m_2 = \left(1 + \frac{1}{\beta}\right)^{1/2}}.$$ (3.14)

So with $\epsilon = 1/100$, the optimum ratio $m_1/m_2 \simeq 10.05$, showing that to maximise the final speed given to the satellite, the first stage must be about ten times larger than the second. Using this new ratio, and with $\epsilon = 0.8$, $c = 3.0$ km s^{-1} as before, (3.13) now gives

$$v \simeq 7.65 \text{ km s}^{-1}.$$

At last we have reached a value close to the required one. So it appears likely that two stage rockets, provided the stage mass ratios are suitably chosen, should be able to launch satellites into earth orbit. Clearly, further staging will improve the final speed, but at the expense of further complexity and increased costs.

3.4 TORRICELLI'S LAW FOR WATER FLOW

In this section we want you to perform an experiment and then attempt to formulate a model to describe the data. First though, the experiment.

Experiment

Take a transparent plastic bottle (e.g. orange squash bottle) which has vertical sides for at least 10 cm. Make a hole about 2 mm in diameter in the side of the bottle near its base. Fix a scale to the side of the bottle, fill the bottle, uncap it and time the flow of water for every cm lost in height. Repeat the experiment three or four times and take the average of the readings.

In the table below we give actual figures obtained when conducting

Height (cm)		11	10	9	8	7	6	5	4	3	2	1
Time (sec)	first run	7	17	29	42	54	68	84	101	121	147	180
	second run	6	17	28	42	53	68	83	101	121	147	181
	third run	7	18	30	41	54	67	84	102	121	147	179
	fourth run	6	17	28	41	53	68	83	100	120	146	179
	fifth run	7	18	30	41	54	68	84	101	121	146	180
	sixth run	6	17	29	41	54	67	83	101	120	146	179
Mean Time		6.5	17.3	29.0	41.3	53.7	67.7	83.5	101.0	120.7	146.5	179.7

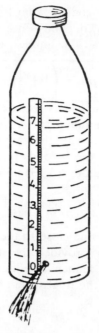

Fig. 3.7—Water Flow Experiment

this experiment six times. Your figures will not be the same unless you are using an identical bottle and same size hole.

We now attempt to formulate a model to explain this data. The important variables are

t — time since water started to flow
h — height of water in bottle above outlet hole
u — volume of water in bottle
v — volume rate of flow through the hole.

If we consider the flow of water through the hole it is clear that u and v are connected by

$$\frac{du}{dt} = -v.$$

Also $u = u_0 + Ah$, where u_0 is the volume of water in the bottle beneath the hole and A is the cross-sectional area. Hence

$$A \frac{dh}{dt} = -v. \qquad (3.15)$$

To proceed further we must consider the form of v. It should be clear from your experiment that v depends on h; in fact as h decreases the volume rate of flow decreases. The simplest form to take is

$$v = ah \qquad (3.16)$$

where a is a positive constant. Thus using (3.16) in (3.15) and writing $\lambda = a/A$ gives the differential equation

$$\boxed{\frac{dh}{dt} = -\lambda h} \; . \qquad (3.17)$$

We have already met this equation in Chapter Two, and the solution is

$$\boxed{h = Ke^{-\lambda h}} \qquad (3.18)$$

where k is an arbitrary constant.

We must now test the quantitative prediction (3.18) against the data determined from the experiments. Taking logs of equation (3.18) gives

$$\ln h = \ln K - \lambda t,$$

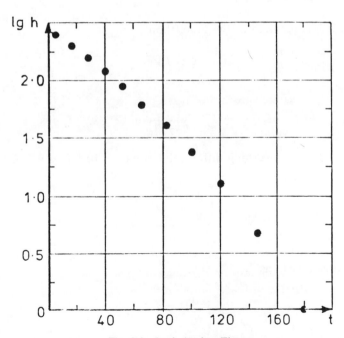

Fig. 3.8—Lg h Against Time t

which indicates that a plot of ln h against t should result in a *straight line*. The experimental results are plotted in Fig. 3.8.

Unfortunately it is clear that the data points would be fitted much better by a downward-bending curve than by a straight line; so we cannot have much confidence in our model. We have been once round the modelling loop, but we must go round again with an improved model.

The least reliable assumption made appears to be the functional for $v = v(h)$, equation (3.16). There are two ways to proceed. Either to analyse the data further, and derive an empirical relationship between v and h which fits the data; or to attempt a better understanding of the underlying physical process.

We can readily attempt the second method since there is a relevant physical principle, called **Torricelli's Law**, which we can use. This states that

'for a non-viscous fluid flowing in a jet from a hole in a container, where the area of the hole is much smaller than the area of the free fluid surface, the velocity of the jet at the hole is given by

$$\boxed{V^2 = 2gh} \tag{3.19}$$

where h is the height of fluid surface above the hole and g the acceleration due to gravity.'

Now if k is the area of the hole, then

$$v = kV$$

and so

$$v = k(2gh)^{1/2}$$

Using (3.15), the governing differential equation becomes

$$\boxed{\frac{dh}{dt} = -\mu h^{1/2}} \tag{3.20}$$

where $\mu = k(2g)^{1/2}/A$. This is a variables separable differential equation which we can write as

$$\int \frac{dh}{h^{1/2}} = \int -\mu \, dt$$

i.e.

$$2h^{1/2} = -\mu t + B$$

$$\boxed{h^{1/2} = -\frac{\mu t}{2} + \frac{B}{2}}.$$ (3.21)

So the theoretical prediction now is that plotting $h^{1/2}$ against t should produce a *straight line*. The plot of experimental data is shown in Fig. 3.9.

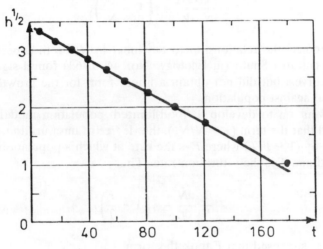

Fig. 3.9—$h^{1/2}$ Against Time t

The straight line agreement is good except for small h. Here the water is no longer issuing as a jet so it is no surprise that the prediction is no longer accurate.

3.5 INHIBITED GROWTH MODELS

In Chapter One we have already met firstly the Malthusian population model

$$\frac{dN}{dt} = \gamma N$$ (3.22)

which predicts unlimited growth (see Fig. 1.1) and the Verhulst model

$$\frac{dN}{dt} = \gamma N - \frac{N^2}{N_\infty}$$ (3.23)

which predicts sigmoid type growth (see Fig. 1.2). Although this second model has been particularly successful in describing some populations there are many for which it is not suitable, and we now look at one of the modifications which has been proposed, namely a food limited population model.

From (3.23) the predicted growth rate per individual is given by

$$\frac{1}{N}\frac{dN}{dt} = \gamma\left(1 - \frac{N}{N_\infty}\right)$$

which predicts the relationship to be linear in N. From experiments on bacteria cultures Smith (in Ecology, No. 44, 1963) found sigmoid type growth curves, but did not obtain a linear form for the growth rate per individual against population.

He went on to develop a 'food-limited' population model. He first supposed that the term $(1 - N/N_\infty)$, the degree of unsaturation, should be replaced by $(1 - F/T)$ where F is the rate at which a population of size N uses food and T is the saturation rate. Thus

$$\frac{1}{N}\frac{dN}{dt} = \gamma\left(1 - \frac{F}{T}\right). \tag{3.24}$$

He further supposed that F took the form

$$F = \lambda N + \mu\frac{dN}{dt} \tag{3.25}$$

here λ and μ are constants. At saturation $F = T$ and $N = N_\infty$ so that we get $T = \lambda N_\infty$; and using this in (3.24) gives

$$\frac{1}{N}\frac{dN}{dt} = \gamma\left(1 - \frac{N}{N_\infty} - \frac{\mu}{\lambda N_\infty}\cdot\frac{dN}{dt}\right).$$

Writing $v = \gamma\mu/\lambda$ gives

$$\boxed{\frac{1}{N}\frac{dN}{dt} = \gamma\,\frac{(N_\infty - N)}{(N_\infty + vN)}.} \tag{3.26}$$

This is the governing differential equation of the model, which no longer predicts a linear growth rate per individual with population size N.

To solve (3.26) we must separate the variables, giving

$$\int \frac{(N_x + vN)}{N(N_x - N)}\, dN = \int \gamma \, dt$$

and partial fractioning the integrand,

$$\int \left[\frac{1}{N} + \frac{(1 + v)}{(N_x - N)} \right] dN = \gamma t + A;$$

$$\ln N - (1 + v) \ln(N_x - N) = \gamma t + A.$$

Thus

$$\boxed{N/(N_x - N)^{1+v} = k e^{\gamma t}} \qquad (3.27)$$

where $k = e^A$ is the constant of integration.

Although we can still obtain a form of the solution, namely (3.27), because of the complexity of the model it is not possible to solve explicitly for N. On the other hand numerical methods can be used to find precise solutions. A typical solution is sketched in Fig. 3.10, and as expected it follows the sigmoid type shape. Smith used this model to describe the growth of bacteria cultures and the model fitted his available data accurately.

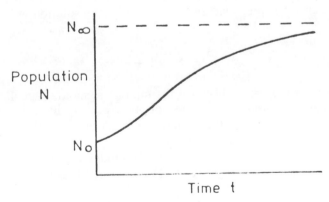

Fig. 3.10—Population Growth for Food Limited Population Model

3.6 THE SPREAD OF TECHNOLOGICAL INNOVATIONS

Once an innovation has been introduced by one firm, how soon do others adopt it? What are the factors which determine how rapidly the innovation spreads? These are questions of vital importance to economists, sociologists and ultimately governments! In this section we construct a model of the spread of a new innovation among farmers.

Suppose that the innovation is introduced into a community of N farmers at time $t = 0$, and let $x(t)$ define the number of farmers who have adopted the innovation at time t. Clearly $x(t)$ has integer values, but we approximate to it as a continuous function of time. Now a farmer will generally adopt the innovation only after he has been told about it by a farmer who already uses it. Hence we assume that the number of farmers, δx, who adopt the innovation in a small time interval, δt, is proportional to

    (i) the number of farmers $x(t)$ who have already adopted;
   (ii) the number of farmers $N - x(t)$ who have not adopted.

Thus

$$\delta x = ax(N - x)\delta t$$

where a is a positive constant. Dividing by δt, and letting $\delta t \to 0$ we obtain the differential equation

$$\boxed{\frac{dx}{dt} = ax(N - x)}. \tag{3.28}$$

This is of the same form as the Verhulst's 'logistic' equation of population growth met in Chapter One and we can write the solution as

$$x = \frac{Ne^{aNt}}{N - 1 + e^{aNt}} \tag{3.29}$$

assuming that $x(0) = 1$. This gives the familiar sigmoid shape illustrated in Fig. 1.2, and so predicts that the adoption process increases up to the value $N/2$, when 50% of the farmers have adopted the innovation, after which the adoption process slows down.

Now we must test the predictions of the model with some actual data on innovations in farming. Figures (3.11) and (3.12) illustrate

    (i) cumulative number of farmers in Iowa during 1944–55 who adopted 2, 4–D weed spray;
   (ii) cumulative percentage of corn acreage in hybrid corn in three American states from 1934–58.

These curves follow sigmoid type shapes and so offer some support to

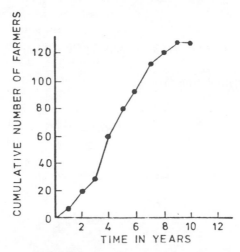

Fig. 3.11—Number of Farmers who Adopted Weed Spray in Iowa

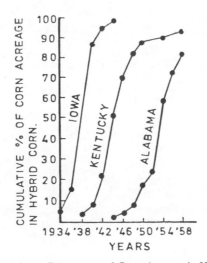

Fig. 3.12—Cumulative Percentage of Corn Acreage in Hybrid Corn in Three American States

the model. One source of any discrepancy might be the assumption that a farmer only learns of an innovation from another farmer. He could easily hear of it through advertising and this could well play a significant part in the adoption process, particularly in its early stages. Suppose that in the small time δt, the number of farmers being influenced through the mass media is proportional to the number of farmers who have not adopted the innovation, i.e. $b(N - x)\,\delta t$ where b is a positive constant. The governing equation is now

$$\delta x = ax(N - x)\delta t + b(N - x)\,\delta t$$

giving the differential equation

$$\boxed{\frac{dx}{dt} = (ax + b)(N - x)}\,.$$ (3.30)

This is again a variables separable equation which we can readily solve, since

$$\int \frac{dx}{(x + c)(N - x)} = \int a\,dt$$

where $c = b/a$. Thus

$$\frac{1}{(N + c)} \int \left[\frac{1}{(x + c)} + \frac{1}{(N - x)} \right] dx = at + A$$

i.e. $$\ln \left[\frac{x + c}{N - x} \right] = (N + c)(at + A),$$

and writing the arbitrary constant as $\ln B = (N + c)A$, we have

$$\frac{x + c}{N - x} = Be^{(N + c)at}\,.$$

Solving for x, and with the initial condition $x(0) = 1$, gives

$$\boxed{x = \frac{(a + b)Ne^{(aN + b)t} - b(N - 1)}{[(N - 1)a + (a + b)e^{(aN + b)t}]}}\,.$$ (3.31)

We again have a sigmoid type growth, and the fit with the data described earlier is good.

The same models as described above have also been used to model the spread of innovations in the coal industry, iron and steel, brewing and railways.

EXERCISES

1. **Reaction to Stimulus**
   (a) Loewenstein Equation
       The differential equation for this model is

$$\frac{dR}{dS} = \frac{R}{(1 + \alpha S)S}$$

   where S is the stimulus and R the reaction, α is a positive constant. Determine R as a function of S and interpret the solution.

   (b) Extension of Steven's Model
       Consider the model described by the equation

$$\frac{dR}{dS} = \frac{cR^{\alpha}}{S^{\beta}}$$

   where c, α and β are positive constants. Solve for R as a function of S and show that both Steven's Law and the Weber-Fechner Law are special cases of this model.

2. **Rocket Flight**
   A single stage rocket expels its propellant at a constant rate k. Assuming constant gravity is the only external force acting, show that the equation of motion is

$$(P + m_0 - kt)\frac{dv}{dt} = ck - (P + m_0 - kt)g$$

   where v is the rocket's speed, c the speed of the rocket propellant relative to the rocket, P the payload mass, and m_0 the initial rocket mass. If the rocket burn is continuous, show that the burn time is $\epsilon m_0/k$ and deduce that the find speed given to the payload is

$$V = -c \ln\left[1 - \frac{\epsilon m_0}{(m_0 + P)}\right] - \frac{g\epsilon m_0}{k},$$

   where $1 - \epsilon$ is the structural factor of the rocket. Estimate the percentage reduction in the predicted final speed due to the inclusion of the gravity term if $\epsilon = 0.8$, $P/m_0 = 1/100$, $c = 3.0$ km s^{-1}, $M = 10^5$ kg, and $k = 5 \times 10^3$ kg s^{-1}.

Also find an expression for the height reached by the rocket during the burn and estimate its value using the data above.

3. **Water Flow**

Using the notation of 3.4, assume that

$$\frac{dh}{dt} = -\lambda h^{\alpha}, \ 0 < \alpha < 1.$$

Deduce the predicted form for h. What quantities give a predicted straight line graph? For $\alpha = 1/3$, test the model against the experimental data found in 3.4.

4. **Inhibited Population Growth**

Investigate the inhibited population model described by the differential equation

$$\frac{dN}{dt} = \gamma N \left\{ 1 - \left[\frac{N}{N_{\infty}} \right]^{\alpha - 1} \right\}$$

where $\alpha > 1$. Solve for N and sketch typical solutions.

5. **Seasonal Growth**

A model of seasonal growth is given by

$$\frac{dx}{dt} = rx \, \cos(wt)$$

where r and w are constants. Illustrate the behaviour of the solution x of this equation.

6. **Chemical Reactions**

(i) In a chemical reaction, two substances, C_1 and C_2, combine in equal amounts to produce a compound, C_3. Suppose that a and b are the initial concentrations (at $t = 0$) of C_1 and C_2. Define $x(t)$ to be the concentration of C_3 at time t. The rate of increase of the concentration of C_3 is $dx/dt = r(a - x)(b - x)$, where r is a positive constant.

(a) If $x(0) = 0$, determine the concentration of C_3 as a function of time for $t > 0$.

(b) If $a = 10$ and $b = 15$ (in appropriate units), determine the limiting concentration of C_3.

(ii) In some chemical reactions, certain products may catalyze their own formation. If $x(t)$ is the amount of such a product at time t, a possible model for such a reaction is given by the differential equation $dx/dt = rx(c - x)$, where r and c are positive constants. In this model, the reaction is completed when $x = c$, presumably because one of the chemicals of the reaction is used up.

   (a) Determine the general solution in terms of the constants r, c, and $x(0)$.
   (b) For $r = 1$, $c = 100$, and $x(0) = 20$, draw a graph of $x(t)$ for $t > 0$.

7. **Cell Growth**
   Nutrients flow across cell walls to determine the growth, survival, and reproduction of the cells. This suggests that, during the early stages of a cell's growth, the rate of increase of the weight of the cell will be proportional to its surface area. If the shape and density of the cell do not change during growth, the weight $x(t)$ of the cell at time t will be proportional to the square of a radius.

   (a) Verify that $x(t)$ satisfies the first order equation $dx/dt = cx^{2/3}$ during the early stages of growth (c is a positive constant).
   (b) In terms of the constant c and the initial weight $x(0)$, determine the weight $x(t)$ of the cell at time t.
   (c) If $c = 3$ and $x(0) = 1$ (in appropriate units), determine the length of time for the weight of the cell to double.

8. **Spread of Epidemics**
   In a model of epidemics, a single infected individual is introduced into a community containing n individuals susceptible to the disease. Define $x(t)$ to be the number of uninfected individuals in the population at time t. If we assume that the infection spreads to all those susceptible, then $x(t)$ decreases from its initial value $x(0) = n$ to zero. A possible equation for $x(t)$ is $dx/dt = -rx(n + 1 - x)$, where r is a positive constant which measures the rate of infection. Determine the solution of this first order equation. When is the infection rate a maximum?

Chapter **4**

Linear First Order Differential Equations

4.1 INTRODUCTION

In this chapter we shall analyse models which lead to linear first order differential equations of the form

$$\boxed{\frac{dy}{dx} + P(x)y = Q(x)} \tag{4.1}$$

where P and Q are functions of x. We can solve differential equations of this form by using the integrating factor method. This involves first evaluating $e^{\int P(x)dx}$, which is called the integrating factor, and then multiplying throughout (4.1) by this function. This gives

$$e^{\int P(x)\,dx}\frac{dy}{dx} + e^{\int P(x)\,dx}P(x)y = e^{\int P(x)\,dx}Q(x).$$

and we note that we can rewrite the left hand side as

$$\frac{d}{dx}\left(e^{\int P(x)dx}y\right) = e^{\int P(x)dx}Q(x).$$

Integrating both sides of this equation, we obtain

$$e^{\int P(x)\,dx}y = \int e^{\int P(x)\,dx}Q(x)\,dx + c,$$

where c is the constant of integration. Thus we can write the solution as

$$\boxed{y = e^{-\int P(x)\,dx}\int e^{\int P(x)\,dx}Q(x)\,dx = ce^{-\int P(x)\,dx}} \; . \tag{4.2}$$

Example

Solve the linear first order differential equation

$$\frac{dy}{dx} + xy = 2x.$$

Solution

Here $P(x) = x$ and $Q(x) = 2x$, so that the integrating factor is $e^{\int x\,dx} = e^{x^2/2}$. Multiplying throughout by $e^{x^2/2}$ gives

$$e^{x^2/2}\frac{dy}{dx} + xe^{x^2/2}y = 2xe^{x^2/2};$$

i.e.

$$\frac{d}{dx}(e^{x^2/2}y) = 2xe^{x^2/2}.$$

Integrating

$$e^{x^2/2}y = \int 2xe^{x^2/2}\,dx + c$$

$$= 2e^{x^2/2} + c.$$

Thus

$$y = 2 + ce^{-x^2/2}.$$

Note

It is of interest to note the form of this solution. The first part of the solution, 2, is a particular solution of the given differential equation whilst $ce^{-x^2/2}$ is the general solution of the associated homogeneous equation

$$\frac{dy}{dx} + xy = 0.$$

4.2 SALES RESPONSE TO ADVERTISING

It is obviously of importance for both advertising agents and their clients to be able to estimate the effectiveness of any proposed advertising campaign. The following model is based on work by Vidale and Wolfe. We first present some of the data on which they based their model, and introduce the parameters of the model.

In the absence of any sort of promotion, sales tend to decrease.

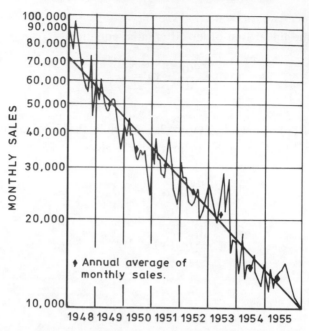

Fig. 4.1—Sales History of Unpromoted Product

Figures 4.1 and 4.2 illustrate the sales history of two products, both of which received no promotion. The second example shows a clear seasonal effect. The data has been plotted on a semi-logarithmic scale.

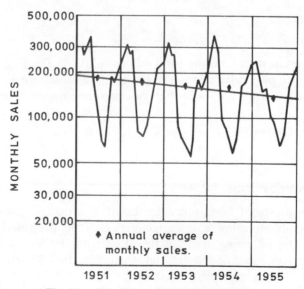

Fig. 4.2—Sales History with Seasonal Effect

   In both cases it appears that the steady decrease in sales on this scale is approximately linear, i.e.

$$\ln S = -\lambda t + \mu,$$

where S is the sales rate, t time and λ and μ constants. Thus

$$\frac{dS}{dt} = -\lambda S, \tag{4.3}$$

when there is no advertising.

   The concept of saturation level is illustrated in Fig. 4.3. This product

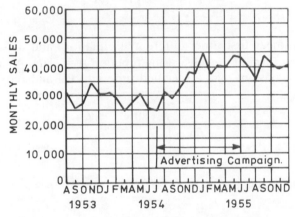

Fig. 4.3—Sales History Illustrating Saturation Sales Level

was promoted continuously for one year by weekly newspaper advertisements. The first six months showed a 30% sales rise while the next six months showed little increase. The additional advertising may have helped maintain the sales at the new rate, but from observations of the sales history before advertising, the decay rate is small. So we conclude that the campaign could have been considerably shorter but still as effective.

   We now formulate the model in mathematical terms. If $A = A(t)$ is the advertising rate, then from (4.3), if $A \equiv 0$,

$$\frac{dS}{dt} = -\lambda S.$$

Now if $A \neq 0$, we assume that the increase in the sales rate is proportional to the advertising rate, A, and also to the degree of unsaturation of the market, namely $(M - S)/M$, where M is the saturation level of the product. Thus M is the practical limit of sales that can be generated, and $(M - S)/M$ is a measure of the market share which has still not bought the product. Combining these assumptions leads to the differential equation

$$\frac{dS}{dt} = rA\,\frac{(M - S)}{M} - \lambda S,$$

where r is a constant, and rearranging gives

$$\boxed{\frac{dS}{dt} + \left(\frac{rA}{M} + \lambda\right)S = rA}\,. \tag{4.4}$$

This is a linear first order differential equation, and its solution will depend on the form of the advertising function $A = A(t)$. As an example, suppose A is constant over a specified time interval and zero thereafter; i.e.

$$A(t) = \begin{cases} \bar{A}, & 0 < t < T \\ 0 & t > T \end{cases} \tag{4.5}$$

and that initially (i.e. at $t = 0$), $S = S_0$. Then for $0 < t < T$,

$$\frac{dS}{dt} + \left(\frac{r\bar{A}}{M} + \lambda\right)S = r\bar{A}$$

and writing $b = r\bar{A}/M + \lambda$, we have integrating factor $e^{\int b\,dt} = e^{bt}$. Thus

$$e^{bt}\frac{dS}{dt} + e^{bt}bS = e^{bt}r\bar{A}$$

i.e.
$$\frac{d}{dt}(e^{bt}S) = e^{bt}r\bar{A}$$

$$e^{bt}S = \int e^{bt}r\bar{A}\,dt = r\bar{A}\int e^{bt}\,dt$$

$$= r\bar{A}\,e^{bt}/b + c$$

where c is the constant of integration. Hence

$$S = r\bar{A}/b + ce^{-bt}$$

for $0 < t < T$. Now at $t = 0$, $S = S_0$ giving

$$S_0 = r\bar{A}/b + c;$$

thus $c = S_0 - r\bar{A}/b$ and for $0 < t < T$

$$S(t) = r\bar{A}/b + (S_0 - r\bar{A}/b)e^{-bt}. \tag{4.6}$$

Now for $t > T$, $A = 0$, and from (4.4) we have

$$\frac{dS}{dt} - \lambda S = 0$$

which has solution $S = ke^{-\lambda t}$, where k is a constant. At $t = T$, $S = S_T$, say, so that $S_T = ke^{-\lambda T}$. Hence for $t > T$,

$$S(t) = S_T e^{-\lambda(t-T)} \tag{4.7}$$

and from (4.6), S_T has the value

$$S_T = r\bar{A}/b + (S_0 - r\bar{A}/b)e^{-bt}.$$

Thus combining (4.6) and (4.7) and substituting for b, we finally

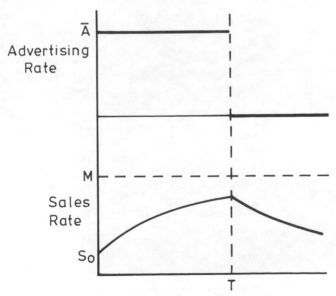

Fig. 4.4—Estimated Sales Response to Advertising

obtain the predicted sales as

$$S(t) = \begin{cases} S_0 e^{-(\lambda + r\bar{A}/M)t} + \dfrac{r\bar{A}}{(\lambda + r\bar{A}/M)} (1 - e^{-(\lambda + r\bar{A}/M)t}) \\ S_T e^{-\lambda(t-T)} \end{cases}$$

A typical solution is sketched in Fig. 4.4.

It can be seen that the rate of increase of sales is most rapid initially, but as saturation is neared the rate decreases. However, when advertising finishes at time T, the sales rate decays exponentially.

Having formulated the model it can now be used to test the

Fig. 4.5—'Woman Taken in Adultery'

effectiveness of advertising campaigns; for example to compare a long, steady campaign as above with a short, intense campaign (see Exercise 1).

4.3 ART FORGERIES

After the liberation of Belgium in World War II, an extensive search was started for the Nazi collaborators. Numerous works of art were sold to the Germans, and a third rate Dutch painter, H. A. Van Meegeran was arrested in May 1945 for collaborating with the enemy in selling Goering a 17th century Vermeer painting, 'Woman Taken in Adultery'.

On July 12, 1945, Van Meegeran surprised the art world by announcing from his prison cell that he was the painter of 'Woman Taken in Adultery', and also of the very famous and beautiful 'Disciples at Emmaus' as well as four other presumed Vermeers and two de Hooghs.

His claims were not at first taken seriously, so whilst in prison he

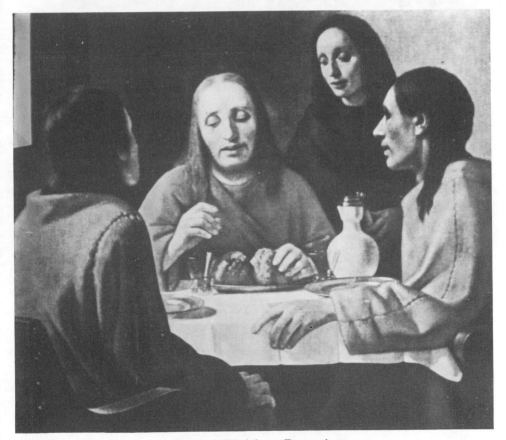

Fig. 4.6—'Disciples at Emmaus'

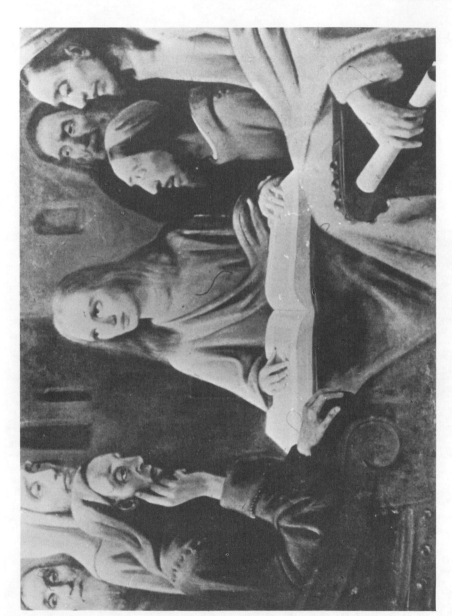

Fig. 4.7—'Jesus Amongst the Doctors'

began to forge the Vermeer painting 'Jesus Amongst the Doctors' in order to show just how good a forger he was. When the work was nearing completion he learned that the charge of collaborating was to be changed to that of forgery, and so he refused to 'age' the painting.

So in order to settle the question of whether the paintings were forgeries, an international panel of distinguished art historians, physicists and chemists was appointed. After some months of work the panel unanimously concluded that the paintings were indeed forgeries. The main reasons for their conclusion were

(i)   the paintings tended to resist water and ethyl alcohol like 17th century paintings, but also resisted strong alkalis and acids, very unlike 17th century paintings;

(ii) evidence was found of how Van Meegeran had tried to produce the 'age' crackle of an old painting;

(iii) the colour cobalt blue, which was not known in the 17th century, was in two of the paintings;

(iv) the medium used in the paintings was an artificial resin of the phenolformaldehyde group, first discovered at the end of the 19th century.

On this evidence, Van Meegeran was convicted on October 12, 1947 and sentenced to one year in prison, but on December 30, 1947 he suffered a heart attack and died.

Despite this evidence, many art experts did not believe that the famed 'Disciples at Emmaus' was a forgery. In fact it was certified by a noted art historian, A. Bredius, and bought by the Rembrandt Society for $170,000. A thoroughly scientific and conclusive proof that 'Disciples at Emmaus' was a forgery was eventually achieved in 1967 by scientists at Carnegie Mellon University. We now describe the basis of their method.

White lead (Pb^{210}) is a radioactive substance with a half life of 22 years (see §2.3 for details of radioactive decay). It is a pigment of major importance in paintings, and it is manufactured from ores which contain uranium and elements to which uranium decays. One of these elements is Radium 226 (Ra^{226}) which has a half life of 1600 years, and decays to Pb^{210}. While still part of the ore the amount of Ra^{226} decaying to Pb^{210} is equal to the amount of Pb^{210} disintegrating per unit time, i.e. Pb^{210} and Ra^{226} are in a 'radioactive equilibrium'.

In the manufacture of the pigment though, the radium and most of its descendants are removed. The Pb^{210} begins to decay and continues to do so until it reaches an 'equilibrium' with the smaller amount of Ra that survived the chemical process.

Let $y(t)$ = amount of Pb^{210} per gm of ordinary lead at time t,

$y(t_0) = y_0$, t_0 being the time of manufacture,
$r(t)$ = number of disintegrations of Ra^{226} per minute per gm of ordinary lead.

Then we use the differential equation model for radioactive decay; i.e.

$$\frac{dy}{dt} = -\lambda y + r(t),$$

(4.8)

λ being the decay constant for Pb^{210},

Equation (4.8) is a linear first order differential equation with integrating factor

$$e^{\int \lambda \, dt} = e^{\lambda t}$$

Thus

$$e^{\lambda t}\frac{dy}{dt} + e^{\lambda t}\lambda y = e^{\lambda t}r(t)$$

i.e.

$$\frac{d}{dt}(e^{\lambda t}y) = e^{\lambda t}r(t).$$

Integrating,

$$e^{\lambda t}y = \int e^{\lambda t}r(t)\,dt + A$$

i.e.

$$y = e^{-\lambda t}\int e^{\lambda t}r(t)\,dt + Ae^{-\lambda t}$$

(4.9)

where A is the constant of integration.
Now if r is assumed constant, then

$$y(t) = e^{-\lambda t}re^{\lambda t}/\lambda + Ae^{-\lambda t}$$

$$= r/\lambda + Ae^{-\lambda t};$$

and at $t = t_0$,

$$y(t_0) = y_0 = r/\lambda + Ae^{-\lambda t_0}.$$

Hence

$$A = e^{\lambda t_0}(y_0 - r/\lambda)$$

and so

$$\boxed{y(t) = r/\lambda + (y_0 - r/\lambda)e^{-\lambda(t-t_0)}}.$$
(4.10)

Is it reasonable to take r as constant? The half life of Ra^{226} is 1600 years, and we are dealing with paintings of about 400 years old. So very little Ra^{226} has decayed in 400 years and so the number of disintegrations of Ra^{226} can be approximated by a constant. We can now use equation (4.8), for we can measure both y and r today, λ is known, and we want to know $(t - t_0)$. Unfortunately we still do not know y_0.

Since the original quantity of Pb^{210} was in equilibrium with the larger quantity of radium in the ore from which the pigment was manufactured, then

$$\lambda y_0 = R,$$
(4.11)

where R denotes the number of disintegrations of Ra^{226} per minute per gm of ordinary lead. But measurements for R from a variety of ores over the earth's surface gives values in the range 0–200. Hence (λy_0) is in this range; but this is not sufficient to obtain an estimate of the date of origin. But we can still make progress, for suppose the painting in question was *not* a forgery but was painted about 300 years ago. Then with $t - t_0 = 300$ in equation (4.10), we can find λy_0, i.e.

$$\lambda y_0 = \lambda y e^{300\lambda} - r(e^{300\lambda} - 1)$$
(4.12)

and $\lambda = 3.151 \times 10^{-2}$. We would expect this number to be in the range 0–200. So any painting giving a value of λy_0 much greater than 200 will almost certainly be a forgery.

For the 'Disciples at Emmaus' painting, (Fig. 4.6)

$\lambda y = 8.5$ (actually polonium 210 is measured, which is approximately the value of lead 210)

$r = 0.8$

and equation (4.12) gives

$$\lambda y_0 = 98,147.$$

This is unacceptably large for a genuine 17th century painting, and so we conclude that 'Disciples at Emmaus' is indeed a forgery.

4.4 ELECTRIC CIRCUITS

In this section we consider simple electric circuits which contain a resistor, an inductor or capacitor in series with a source of electromotive force (emf). Two such circuits are illustrated in Fig. 4.8.

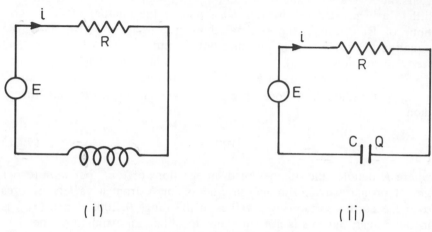

(i)   (ii)

Fig. 4.8—Simple Electrical Circuits

The dynamics of electric circuits are based on the following assumptions:

1. An electromotive force (emf) E (measured in volts), e.g. battery, drives an electric charge Q (measured in coulombs) and produces a current I (measured in amperes). The current and charge are related by

$$I = \frac{dQ}{dt}.$$
(4.13)

2. A resistor of resistance R (measured in ohms) opposes the current, dissipating the energy in the form of heat. The drop in voltage produced is given by Ohm's Law as

$$E_R = RI.$$
(4.14)

3. An inductor of inductance L (measured in henrys) produces a drop in voltage given by

$$E_L = L \frac{dI}{dt}.$$
(4.15)

4. A capacitor of capacitance C (measured in farads) stores charge, and so resists the flow of charge, producing a voltage drop

$$E_C = \frac{Q}{C}. \tag{4.16}$$

The quantities R, L and C are constants associated with the components in the circuit, whilst E may be a constant or function of time. The voltage drops in a circuit are related by Kirchhoff's voltage law: 'The algebraic sum of all voltage drops around a closed circuit is zero'.

Hence for circuit (i) illustrated in Fig. 4.8,

$$E_R + E_L - E = 0;$$

and using (4.14) and (4.15) gives

$$RI + L\frac{dI}{dt} = E$$

i.e.

$$\frac{dI}{dt} + \frac{R}{L}I = \frac{E}{L}.$$

This is a first order linear differential equation with integrating factor $e^{\int R/L \, dt} = e^{Rt/L}$. Thus we write

$$e^{Rt/L}\frac{dI}{dt} + \frac{R}{L}e^{Rt/L}I = \frac{E}{L}e^{Rt/L}$$

i.e.

$$\frac{d}{dt}\left(e^{Rt/L}I\right) = \frac{E}{L}e^{Rt/L}.$$

Integrating,

$$e^{Rt/L}I = \int \frac{E}{L}e^{Rt/L}\,dt + A$$

i.e

$$\boxed{I(t) = e^{-Rt/L}\int \frac{E(t)}{L}e^{Rt/L}\,dt + Ae^{-Rt/L}} \tag{4.17}$$

where A is the constant of integration.

As an example, if $E(t) = E$, constant, then

$$I(t) = E/R + Ae^{-Rt/L},$$

and if $I(0) = I_0$, the current is given by

$$I(t) = E/R + (I_0 - E/R)e^{-Rt/L}. \tag{4.18}$$

The form of this solution is illustrated in Fig. 4.9.

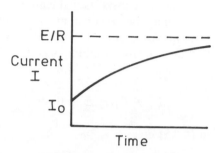

Fig. 4.9—Changes in Current

As can be seen, the current tends to a steady state as time increases.

The form of Kirchhoff's law for circuit (ii) in Fig. 4.8 is

$$E_R + E_C - E = 0;$$

using (4.14) and (4.16) gives

$$R\frac{dQ}{dt} + \frac{Q}{C} = E,$$

since $I = dQ/dt$. We leave this example as an exercise.

4.5 EXPLOITED FISH POPULATIONS

Uncontrolled fishing at sea would eventually seriously deplete world fish stocks and so in the past decade countries and international commissions have imposed new fishing boundaries and defined limits to fishing catches. International agreements are usually based on several types of controls; e.g.

  (i) net size
 (ii) catch size or trawler fleet size
(iii) periods of fishing

In order to predict the effect of such controls, mathematical models must be used.

We start the modelling process by considering the growth of an individual fish. If $w = w(t)$ denotes the weight of a fish, the von Bertalanffy growth model is

$$\frac{dw}{dt} = \alpha w^{2/3} - \beta w \quad ; \tag{4.19}$$

the first term is the increase in weight due to nutrients and is taken as proportional to a surface area; the second term is the rate of loss of weight due to respiration and is taken as proportional to the weight.

This is a *non-linear* differential equation since it contains the $w^{2/3}$ term. It is, in fact, a special case of Bernoulli's differential equation, which takes the form

$$\frac{dy}{dx} + f(x)y = g(x)y^n \tag{4.20}$$

where f and g are functions of x. This type of differential equation can be reduced to a linear first order equation by using the substitution $y^{1-n} = v$. Thus for (4.19),

$$\frac{dw}{dt} + \beta w = \alpha w^{2/3}$$

we make the substitution

$$v = w^{1-2/3} = w^{1/3}.$$

Then

$$\frac{dv}{dt} = \frac{1}{3} w^{-2/3} \frac{dw}{dt} = \frac{1}{3} w^{-2/3}(\alpha w^{2/3} - \beta w)$$

$$= \frac{\alpha}{3} - \frac{\beta}{3} v$$

i.e.

$$\frac{dv}{dt} + \frac{\beta}{3} v = \frac{\alpha}{3} \quad . \tag{4.21}$$

This is a linear first order differential equation with integrating factor

$e^{\int(\beta/3)\,dt} = e^{\beta t/3}$. Thus

$$e^{\beta t/3}\frac{dv}{dt} + \frac{\beta}{3}e^{\beta t/3}v = e^{\beta t/3}\frac{\alpha}{3},$$

i.e.

$$\frac{d}{dt}(e^{\beta t/3}v) = e^{\beta t/3}\alpha/3;$$

and integrating,

$$e^{\beta t/3}v = \int e^{\beta t/3}(\alpha/3)\,dt = (\alpha/\beta)e^{\beta t/3} + A,$$

where A is the constant of integration. Rearranging,

$$v = (\alpha/\beta) + Ae^{-\beta t/3},$$

so that

$$w = \left(\frac{\alpha}{\beta}\right)^3\left(1 + \frac{A\beta}{\alpha}e^{-\beta t/3}\right)^3.$$

Now if $w \to w_\infty$ as $t \to \infty$, we must have $w_\infty = (\alpha/\beta)^3$; and putting $w = 0$ at $t = 0$ gives $A = -\alpha/\beta$. Hence the predicted growth in fish size is given by

$$\boxed{w = w_\infty[1 - e^{-\gamma t}]^3} \tag{4.22}$$

where $\gamma = \beta/3$. This is illustrated in Fig. 4.10.

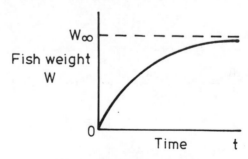

Fig. 4.10—Fish Growth

Let us consider the simplified problem of what happens to an initial group of young fish, say total number n_0. If there is no harvesting, for various other reasons, the number in this group would decline. We will suppose that the total number declines in an exponential manner, i.e.

$$n(t) = n_0 e^{-Rt} \qquad (4.23)$$

where R is a constant. So $n(t)$ is the number of fish which survive to age t years when there is no harvesting. Note as time increases, the size of each individual fish increases according to formula (4.22), so that the total weight of the fish population is increasing even though the total number is decreasing.

We now incorporate into the model the effect of fishing. Fish which are too young (i.e. too small) to be caught can be avoided by using a certain minimum mesh size for the nets; so we can suppose that fish no younger than, say, M years old are caught. The effect of fishing can be modelled by supposing that the decline rate is increased from R to $R + F$, where F is some measure of the fishing effort.

So at $t = M$, $n = n_0 e^{-RM}$, and for $t \geqslant M$,

$$n(t) = n_0 e^{-RM} e^{-(R+F)(t-M)} \qquad (4.24)$$

This is illustrated in Fig. 4.11. The yield is now given by

$$Y = \int_M^\infty w(t) \times (\text{number of fish caught at age } t) \, dt$$

$$= \int_M^\infty w_\infty (1 - e^{-\gamma t})^3 F n_0 e^{-RM} e^{-(R+F)(t-M)} \, dt,$$

where the factor F represents the proportion of the fish alive at age t which are caught. Evaluating this integral gives

$$Y = Y(F, M) \qquad (4.25)$$

i.e. the yield is a function of the controls, F the fishing effort and M the minimum size caught.

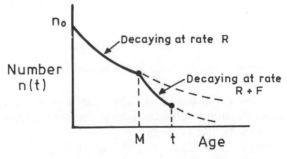

Fig. 4.11—Number of Fish Surviving to Age t Years

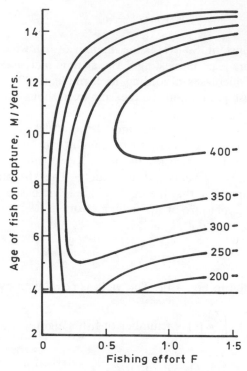

Fig. 4.12—Yield Contours

This model has been applied to North Sea plaice, and Fig. 4.12 illustrates the Y contour for changing F and M. Each curve represents a constant yield, and the yield has a predicted maximum at about $M = 11$, $F = 1.2$. In fact, it is estimated that plaice stocks are actually being exploited at $M = 4$, $F = 0.7$, and it appears as if the total plaice stock is not in decline. But the model above is just concerned with maximising the possible yield from one group of fish, and does not take into account future supply of fish.

4.6 NEOCLASSICAL ECONOMIC GROWTH

In this section we will consider a mathematical model of a simplified economy in which there is no foreign trade and only one homogeneous article is produced. The purpose of such models is to clarify and analyse key problems in economic growth theory, and not to provide accurate descriptions of a particular economy.

We first introduce the output of the product, say $Y = Y(t)$, t being time. We suppose that this output is based on two inputs, namely capital

$K = K(t)$ and the labour force $L = L(t)$. Hence we can write

$$Y = Y(K, L) \qquad (4.26)$$

where F is called the production function. We further suppose that F possesses a 'return to scale' property, which in mathematical terms can be stated as

$$F(\alpha K, \alpha L) = \alpha F(K, L); \qquad (4.27)$$

this means, for example, that a doubling of both the capital and labour force would produce a doubling of output.

Now with $\alpha = 1/L$ in (4.27), and defining output per worker as $y = Y/L$ and capital per worker as $k = K/L$, we have

$$y = \frac{Y}{L} = \frac{1}{L} F(K, L) = F(k, 1).$$

Writing $F(k, 1) = f(k)$, we see that

$$y = f(k). \qquad (4.28)$$

The production function f is usually taken as a strictly concave monotonic increasing function with

$$\lim_{k \to 0} \left(\frac{df}{dk} \right) = \infty, \; \lim_{k \to \infty} \left(\frac{df}{dk} \right) = 0.$$

A typical production function is sketched in Fig. 4.13

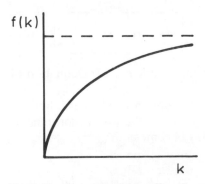

Fig. 4.13—Typical Production Function

Now output from production is either consumed or invested; so we write

$$Y(t) = C(t) + I(t) \qquad (4.29)$$

where C and I are the rates of consumption and investment respectively. Investment is used to augment the capital stock, so that

$$\frac{\mathrm{d}K}{\mathrm{d}t} = I \qquad (4.30)$$

Combining (4.28) and (4.29) with (4.30) we see that

$$y = c(t) + \frac{1}{L}\frac{\mathrm{d}K}{\mathrm{d}t}$$

where $c = C/L$ is the consumption per worker. Also

$$\frac{\mathrm{d}k}{\mathrm{d}t} = \frac{\mathrm{d}}{\mathrm{d}t}\left(\frac{K}{L}\right) = \frac{1}{L}\frac{\mathrm{d}K}{\mathrm{d}t} - \frac{\dot{L}}{L}k,$$

and assuming an exponentially increasing labour force, $L = L_0 e^{\lambda t}$, we see that

$$\frac{\mathrm{d}k}{\mathrm{d}t} = \frac{1}{L}\frac{\mathrm{d}K}{\mathrm{d}t} - \lambda k.$$

Eliminating $(1/L)\mathrm{d}K/\mathrm{d}t$ we arrive at the 'fundamental equation' of neoclassical economic growth

$$\boxed{\frac{\mathrm{d}k}{\mathrm{d}t} = f(k) - \lambda k - c} \ . \qquad (4.31)$$

For a given consumption profile $c = c(t)$, equation (4.31) determines the capital path.

For example, if consumption is taken as a linearly increasing function with time, i.e. $c = at + b$, and the production function is assumed linear, i.e. $f = \mu k$, then (4.31) becomes

$$\frac{\mathrm{d}k}{\mathrm{d}t} - (\mu - \lambda)k = -(at + b). \qquad (4.32)$$

Writing $v = \mu - \lambda$, this is a linear first order differential equation with integrating factor e^{-vt}. Thus

$$\frac{d}{dt}\left(e^{-vt}k\right) = -e^{-vt}(at + b)$$

and integrating

$$
\begin{aligned}
e^{-vt}k &= \int -e^{-vt}(at + b)\, dt \\
&= e^{-vt}(at + b)/v - \int (e^{-vt}/v)\, a\, dt \\
&= e^{-vt}(at + (a/v) + b)/v + A
\end{aligned}
$$

where A is the constant of integration. If $k = k_0$ at $t = 0$, $A = [k_0 - (a/v + b)/v]$ and

$$k(t) = \{at + (a/v) + b + e^{vt}[vk_0 - (a/v + b)]\}/v. \qquad (4.33)$$

The predictions of this model are very much dependent on the relative value of the parameters v, a, b and k_0. If, for example, $\mu > \lambda$ so that $v > 0$, continued capital growth is predicted provided

$$k_0 > (a/v + b)/v = k_c, \quad \text{say}.$$

If the initial capital stock is less than k_c, capital will decrease and the economy is in decline.

This is a simple example of the 'threshold effect' in economics, where a certain level of economic development ($k > k_c$ in this case) is the precondition for further capital accumulation.

The linear production function used in the above example does not conform to all the usual properties of production functions. A better function to explore is the Cobb–Douglas production function

$$f(k) = k^\alpha, \; 0 < \alpha < 1, \qquad (4.34)$$

and this is developed in the exercises.

4.7 POLLUTION OF THE GREAT LAKES

Industrialised nations are facing the problems of water pollution. Once pollution of a river is stopped, it will rapidly clean itself provided the pollution has not caused extreme damage. Lakes are not quite so easy to deal with since a considerable amount of water has to be cleaned. How long will this take; and, for example, how long would it take to clean up the Great Lakes of North America?

   The main cleanup mechanism is the natural process of gradually replacing the water in the lake, provided of course that the pollution has not caused irreversible damage and already 'killed' the lake. Fig. 4.14 shows a map of the Great Lakes. The basic idea behind the model is to regard the flow in the lakes as a perfect mixing problem, ignoring biological action, sedimentation etc. The following assumptions are made:

1. Rainfall and evaporation balance each other;
2. When water enters the lake, perfect mixing occurs, so that pollutants are uniformly distributed;
3. Pollutants are only removed from the lake by outflow.

Fig. 4.14—The Great Lakes

   By these assumptions, the net change in total pollutants during the time interval δt is

$$\delta(VP_l) = (P_i - P_l)(r\delta t)$$

where V is the volume of the lake, P_l is the pollution concentration in the lake, P_i is the pollution concentration in the inflow to the lake, r is the rate of flow. Thus dividing by δt and letting $\delta t \to 0$, we obtain the differential equation

$$\frac{dP_l}{dt} = \frac{(P_i - P_l)r}{v} .$$ (4.35)

This is a linear first order differential equation with integrating factor $e^{\int r/v \, dt} = e^{rt/v}$; i.e.

$$\frac{d}{dt}(e^{rt/v}P_l) = e^{rt/v}P_i r/v.$$

Integrating both sides and evaluating at $t = 0$ and t gives

$$e^{rt/v}P_l(t) - P_l(0) = \int_0^t (e^{rt/v}P_i \, r/v) \, dt$$

i.e.

$$P_l(t) = e^{-rt/v}P_l(0) + e^{-rt/v}(r/v)\int_0^t e^{rt/v}P_i \, dt .$$ (4.36)

The numbers in Fig. 4.14 are Rainey's values for $\tau = v/r$; a value for Huron is not given. Using (4.36) we can determine the effect of various anti-pollution schemes. (Lake Ontario is not included since about 84% of its inflow comes from Lake Erie.) The fastest possible cleanup will occur if all the pollution inflow ceases; i.e. $P_i = 0$, so that $P_l(t) = e^{-rt/v}P_l(0)$ giving

$$t = \tau \log\left(\frac{P_l(0)}{P_l(t)}\right)$$ (4.37)

This formula can be used to find out how long it would take to reduce pollution to a given percentage of its present level. The table below gives the time in years.

Lake	50%	20%	10%	5%
Erie	2	4	6	8
Michigan	21	50	71	92
Superior	131	304	435	566

Fortunately, Lake Superior's pollution is quite low at the present time.

The model is clearly a very simplified one, but nevertheless the figures in the table probably provide rough lower bounds for the cleanup times.

EXERCISES

1. **Sales Response to Advertising**
    (i) Analyse the sales response to the advertising model of §4.2 when the following advertising campaigns are implemented:

$$\text{(a) } A(t) = \alpha t \ (\alpha \text{ positive constant})$$
$$\text{(b) } A(t) = \beta \ (\beta \text{ positive constant}).$$

In both cases take the decay constant as zero, and compare the total sales over a time interval $[0, T]$ when the total advertising budget is fixed; i.e.

$$\int_0^T A(t) \, dt = A_0, \text{ constant.}$$

(ii) Compare a constant advertising campaign, with a campaign consisting of two short intervals of higher intensity.

2. **Art Forgeries**
    Use the model in §4.3 to determine from the data given below whether the following paintings are forged Vermeers:

Picture	Polonium 210 disintegrations	Radium 226 disintegrations
'Washing of Feet'	12.6	0.26
'Woman Reading Music'	10.3	0.3
'Woman Playing Mandolin'	8.2	0.17
'Lace Maker'	1.5	1.4
'Laughing Girl'	5.2	6.0

3. **Electric Circuits**
    (i) Derive the circuit equation for the electric circuit shown in Fig. 4.9 (ii). Deduce that the solution for the electric charge is

$$Q(t) = e^{-t/CR} \left[\frac{1}{R} \int E(t) e^{t/CR} \, dt + c \right].$$

Also find an expression for the current I. Evaluate Q and I if E is a constant.

(ii) Find the electric current for the circuit in Fig. 4.9 (i) when the e.m.f. is periodic, i.e.

$$E = E_0 \cos(wt)$$

where E_0 and w are positive constants.

4. Fishing Yield

Determine the yield function $Y = Y(F, M)$, equation (4.25) in §4.5. For given values of F, show how Newton's method for finding approximate roots of equation can be used to determine M, when the yield Y is a constant.

5. Economic Growth

(i) If consumption is taken as a constant fraction, say μ, and using the Cobb–Douglas Production function, $f = k^\alpha$, $0 < \alpha < 1$, show that the neoclassical economic growth model takes the form

$$\frac{dk}{dt} = k^\alpha - (\lambda + \mu)k.$$

Use the substitution $v = k^{\alpha-1}$ to reduce this equation to a linear first order differential equation. Hence solve for k.

(ii) Determine equilibrium positions for capital, i.e. $dk/dt = 0$ if consumption is constant. Show that there are two positions, one position or none depending on the value of c. If $f = k^\alpha$, show that the maximum consumption position occurs when $\alpha k^{\alpha-1} = \lambda$. This is called the 'golden rule' level of capital. Find the corresponding consumption level.

Chapter 5

Linear Second Order Differential Equations

5.1 INTRODUCTION

In this chapter we are concerned with mathematical models which lead to the differential equation

$$\boxed{\frac{d^2y}{dx^2} + a\frac{dy}{dx} + by = f(x)}.$$

(5.1)

This is a second order linear differential equation; we shall just be concerned with this equation when a and b are constants; the function $f(x)$ is a specified function of x.

The key to solving differential equations of this type is to find one particular solution, say $y_p(x)$, of equation (5.1). We now consider the difference between the general solution $y(x)$ of (5.1) and this particular solution, $y_p(x)$. Let

$$z(x) = y(x) - y_p(x),$$

then

$$\frac{d^2z}{dx^2} + a\frac{dz}{dx} + bz = \frac{d^2y}{dx^2} + a\frac{dy}{dx} + by - \left(\frac{d^2y_p}{dx^2} + a\frac{dy_p}{dx} + by_p\right)$$

$$= f(x) - f(x) = 0$$

since both $y(x)$ and $y_p(x)$ satisfy (5.1). Hence the function $z(x)$ satisfies the associated homogenous equation

$$\frac{d^2y}{dx^2} + a\frac{dy}{dx} + by = 0.$$

(5.2)

We usually write $z(x)$ as $y_c(x)$, and this function is called the complementary function (or kernel). Thus the general solution of (5.1) takes the form

$$y(x) = y_c(x) + y_p(x). \tag{5.3}$$

Example

Find the general solution of

$$\frac{d^2y}{dx^2} + y = x.$$

Solution

The general solution has the form $y(x) = y_p(x) + y_c(x)$, where y_c is the general solution of

$$\frac{d^2y}{dx^2} + y = 0$$

and y_p is one particular solution. Now it is easy to see that

$$y_c(x) = A \sin x + B \cos x,$$

where A and B are arbitrary constants, and it can also be seen that a particular solution is given by $y_p = x$. By (5.3) the general solution is

$$y = A \sin x + B \cos x + x.$$

It should be noted at this stage that the general solution of the linear homogeneous differential equation (5.2) is always of the form

$$\boxed{Ay_1(x) + By_2(x)}, \tag{5.4}$$

where $y_1(x)$, $y_2(x)$ are two linearly independent solutions of (5.2) and A and B are arbitrary constants. Two functions $y_1(x)$, $y_2(x)$ are said to be linearly independent if

$$\alpha_1 y_1(x) + \alpha_2 y_2(x) = 0$$

for constants α_1 and α_2 implies $\alpha_1 = \alpha_2 = 0$. For example, take $y_1 = x$, $y_2 = x^2$, then

$$\alpha_1 x + \alpha_2 x^2 = 0$$

is only satisfied for all x if $\alpha_1 = \alpha_2 = 0$. Thus x and x^2 are linearly independent; whereas x and $2x$ are linearly dependent, since

$$1 . x + \tfrac{1}{2} . 2x = 0$$

for all x.

For the solution of $d^2y/dx^2 + y = 0$, we have $y_1 = \sin x$ and $y_2 = \cos x$, which are linearly independent giving the general solution as

$$y = A \sin x + B \cos x.$$

When the coefficients a, b are constant, it is relatively straightforward to solve (5.2),

$$\boxed{\frac{d^2y}{dx^2} + a \frac{dy}{dx} + by = 0} \; .$$

To find a solution, we try $y = e^{mx}$, where m is a constant to be determined. Then, to satisfy the differential equation, we require

$$e^{mx}(m^2 + am + b) = 0,$$

and since this must be true for all appropriate x, we conclude that

$$m^2 + am + b = 0. \tag{5.5}$$

This is a quadratic equation in m, called the auxiliary equation, and will in general have two solutions, say m_1 and m_2 giving two solutions, $e^{m_1 x}$ and $e^{m_2 x}$ which are linearly independent. Hence the general solution of (5.2) takes the form

$$y(x) = Ae^{m_1 x} + Be^{m_2 x}. \tag{5.6}$$

There are three cases to consider, depending on the sign of '$b^2 - 4ac$', i.e. $a^2 - 4b$.

(i) $a^2 - 4b > 0$ — in this case there will be two real solutions e.g. $d^2y/dx^2 + 3dy/dx + 2y = 0$ has auxiliary equation $m^2 + 3m + 2 = 0$, giving $(m + 2)(m + 1) = 0$ and so $m_1 = -2$, $m_2 = -1$ and its general solution is

$$y = Ae^{-2x} + Be^{-x}.$$

(ii) $a^2 - 4b = 0$ — we now have a repeated root $m = -a/2$ and a second linearly independent solution is given by xe^{mx}; e.g. $d^2y/dx^2 - 4dy/dx + 4y = 0$ has auxiliary equation $m^2 - 4m + 4 = 0$, giving $(m - 2)^2 = 0$. Hence $y_1 = e^{2x}$ is a solution, and it can soon be verified that $y_2 = xe^{2x}$ is also a solution. The general solution is then given by

$$y = Ae^{2x} + Bxe^{2x}.$$

(iii) $a^2 - 4b < 0$ — here we have complex roots, say $m_1 = \alpha + i\beta$, $m_2 = \alpha - i\beta$ so that we can formally write the general solution as

$$y = Ae^{(\alpha+i\beta)x} + Be^{(\alpha-i\beta)x}$$

$$= e^{\alpha x}(Ae^{i\beta x} + Be^{-i\beta x})$$

$$= e^{\alpha x}((A + B)\cos \beta x + i(A - B)\sin \beta x)$$

since $e^{i\theta} = \cos \theta + i \sin \theta$. Writing new arbitrary constants $C = A + B$, $D = i(A - B)$, we have the solution

$$y = e^{\alpha x}(C \cos \beta x + D \sin \beta x) \qquad (5.7)$$

e.g. $d^2y/dx^2 + 2dy/dx + 2y = 0$ has auxiliary equation $m^2 + 2m + 2 = 0$ giving $m = -1 \pm i$. Thus $\alpha = -1$, $\beta = 1$ and we can write the solution as

$$y = e^{-x}(C \cos x + D \sin x).$$

So we have considered all the possible cases when solving (5.2), and in order to solve (5.1) we are now only left with finding one particular solution, y_p, of (5.1). Unfortunately this is not as straightforward as finding y_c, and we will use a 'trial and error' method, although there are more precise techniques available.

5.2 MECHANICAL OSCILLATIONS

We will consider an idealised vibrational system in this section, which is illustrated in Fig. 5.1. A particle of mass m is attached by a spring to a fixed point. It is assumed that the spring obeys Hooke's Law (tension in spring is proportional to its extension), that there is a resistance to motion proportional to the particle's speed, and that there is an external force $F(t)$ per unit mass, applied to the particle.

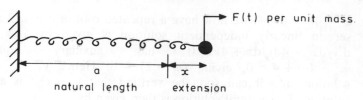

Fig. 5.1—Mechanical System

With the above assumptions, if x is the spring's extension from its natural length a, the equation of motion of the particle of mass m is

$$m\ddot{x} = -T + mF - mk\dot{x} \qquad \text{('dot' means d/d}t)$$

where T is the force in the spring, i.e. $T = \lambda x/a$, λ constant, and k is a constant, which is a measure of the resistance. Thus we have the differential equation

$$\boxed{\ddot{x} + k\dot{x} + \omega^2 x = F(t)} \qquad (5.8)$$

where $\omega^2 = \lambda/am$. We start our analysis with a very special case, namely no resistance ($k = 0$) and no external force ($F = 0$), which gives

$$\boxed{\ddot{x} + \omega^2 x = 0} . \qquad (5.9)$$

This has general solution

$$x = A \cos \omega t + B \sin \omega t. \qquad (5.10)$$

An alternative form of this solution is

$$\boxed{x = a \cos(\omega t + \beta)} . \qquad (5.11)$$

We can see the equivalence of (5.10) and (5.11) by expanding (5.11) to give

$$x = \alpha \cos \omega t \cos \beta - \alpha \sin \omega t \sin \beta$$

which is (5.10) with $\alpha \cos \beta = A$ and $-\alpha \sin \beta = B$. The solution (5.11) predicts that the motion is purely oscillatory with period $2\pi/\omega$. We call this *simple harmonic motion*, and it is illustrated in Fig. 5.2. We call ω the *natural frequency* of the system, and α is called the *amplitude*.

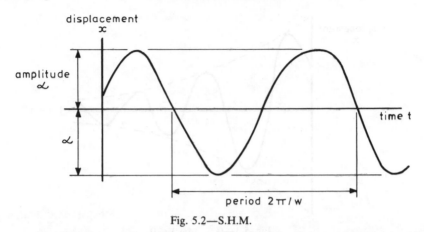

Fig. 5.2—S.H.M.

We now include the resistance, or damping, term and so consider the equation

$$\ddot{x} + k\dot{x} + \omega^2 x = 0 \quad . \tag{5.12}$$

Assuming $x = e^{mx}$, the auxiliary equation is

$$m^2 + km + \omega^2 = 0$$

and so

$$m = [-k \pm (k^2 - 4\omega^2)^{1/2}]/2 = m_1, m_2 \text{ say.}$$

The form of the solution will depend on the sign of $(k^2 - 4\omega^2)$

(i) $k^2 - 4\omega^2 < 0$ – In this case, if $c = (4\omega^2 - k^2)^{1/2}/2$, then

$$m_1 = -k/2 + ic, m_2 = -k/2 - ic$$

and the solution is

$$x = e^{-kt/2} (A \cos ct + B \sin ct).$$

The displacement is illustrated in Fig. 5.3. The motion is oscillatory (as in the case $k = 0$) but the amplitude decreases to zero.

(ii) $k_2 - 4\omega^2 = 0$ — In this case, we have equal roots, $m = -k/2$,

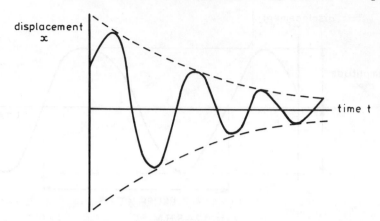

Fig. 5.3—$(k^2 - 4\omega^2) < 0$

and so the solution is

$$x = Ae^{-kt/2} + Bte^{-kt/2}.$$

The displacement is illustrated in Fig. 5.4. The motion is no longer oscillatory, and as in case (i), $x \to 0$ as $t \to \infty$.

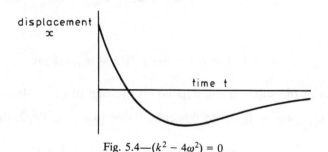

Fig. 5.4—$(k^2 - 4\omega^2) = 0$

(iii) $k^2 - 4\omega^2 > 0$ — In this case, we have two real roots

$$m_1 = [-k + (k^2 - 4\omega^2)^{1/2}]/2$$

$$m_2 = [-k - (k^2 - 4\omega^2)^{1/2}]/2$$

and both are negative. The displacement is illustrated in Fig. 5.5. Again $x \to 0$ as $t \to \infty$.

In all three cases, the important property is that the displacement

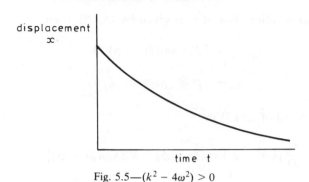

displacement
x

time t

Fig. 5.5—$(k^2 - 4\omega^2) > 0$

tends to zero as time increases. So the effect of resistance is to *dampen* the motion, the dampening becoming more severe as k increases.

We now turn to the general problem, including both resistance and an external force. The motion is now governed by equation (5.8), i.e.

$$\ddot{x} + k\dot{x} + \omega^2 x = F(t), \tag{5.13}$$

and we note that the form of the solution is

$$x = x_p + x_c, \tag{5.14}$$

where x_p is a particular solution (and so depends on the external force F) and x_c is the solution of the associated homogeneous equations (5.12). We have already seen above that $x_c \to 0$ as $t \to \infty$, so x_c is called the *transient*. From (5.14), we see that $x \to x_p$ as $t \to \infty$, and so the vital part of the solution is x_p.

For example, if the external force takes an oscillatory form, i.e.

$$\boxed{F(t) = F_0 \cos \beta t} \tag{5.15}$$

it can be checked that

$$x_p = \frac{F_0}{D} \cos(\beta t - \phi) \tag{5.16}$$

where

$$D = [(\omega^2 - \beta^2)^2 + k^2 \beta^2]^{1/2} \tag{5.17}$$

and

$$\sin \phi = k\beta/D, \quad \cos \phi = (\omega^2 - \beta^2)/D \tag{5.18}$$

is a particular solution. For, if x_p is given by (5.16), then

$$\dot{x}_p = -F_0\beta \sin(\beta t - \phi)/D$$

$$\ddot{x}_p = -F_0\beta^2 \cos(\beta t - \phi)/D$$

and the left hand of (5.13) is

$$\frac{F_0}{D}[(\omega^2 - \beta^2)\cos(\beta t - \phi) - k\beta \sin(\beta t - \phi)]$$

$$= F_0[\cos\phi \cos(\beta t - \phi) - \sin\phi \sin(\beta t - \phi)]$$

$$= F_0 \cos(\phi + \beta t - \phi)$$

$$= F_0 \cos\beta t,$$

which is the right hand side of (5.13) when $F(t) = F_0 \cos\beta t$.

So when the external force is given by $F_0 \cos\beta t$, the displacement $x \to x_p = F_0 \cos(\beta t - \phi)/D$. So the forced oscillation has the same period as the applied force, but with a phase change ϕ and a modified amplitude, $F_0/D = F_0/[(\omega^2 - \beta^2)^2 + k^2\beta^2]^{1/2}$. The forcing function and the response, x_p, are illustrated in Fig. 5.6.

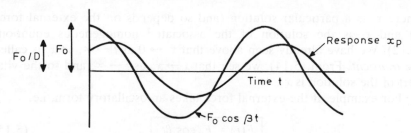

Fig. 5.6—Forcing Function and Response

We can see that the amplitude modification, D, depends not only on the natural frequency, ω, and forcing frequency, β, but also the damping coefficient k. As $k \to 0$, we have

$$x_p \approx \frac{F_0 \cos\beta t}{(\omega^2 - \beta^2)}, \qquad (5.19)$$

which implies the response to the forcing term is oscillatory, with no phase change but with a modified amplitude $F_0/(\omega^2 - \beta^2)$. There is nothing particularly surprising yet, but consider what happens if the forcing frequency β becomes close to the natural frequency ω. The amplitude gets larger, and the solution breaks down when $\omega = \beta$.

To find out what actually happens when $\omega = \beta$ we must solve the differential equation (5.13) with $k = 0$ and $F = F_0 \cos \omega t$, i.e.

$$\ddot{x} + \omega^2 x = F_0 \cos \omega t \qquad . \qquad (5.20)$$

It can soon be shown that

$$x_p = \frac{F_0 t}{2\omega} \sin \omega t \qquad (5.21)$$

and this is illustrated in Fig. 5.7. The characteristics of the response have now changed dramatically. The motion is still oscillatory, but its amplitude, $F_0 t / 2\omega$, now steadily increases without limit. So the displacement will oscillate with ever increasing amplitude.

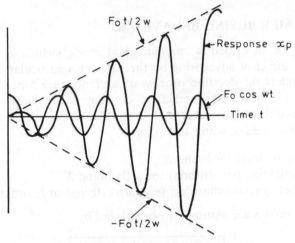

Fig. 5.7—Resonance

This phenomenon is known as *resonance*, and occurs when the forcing and natural frequencies are equal. So for this simplified model, with resistance neglected, as time increases, the displacement forced on the system oscillates with the same period as the forcing function, but with rapidly increasing amplitude. For structures this can have dire consequences, so that engineers and scientists must guard against resonance.

Two notable cases when resonance has proved disastrous are

(i) *Broughton Suspension Bridge* (near Manchester, 1831)
A column of soldiers marched over the bridge, thus setting up a

periodic force, which happened to have a frequency approximately equal to one of the natural frequencies of the bridge. The bridge collapsed, and so soldiers no longer march in step across bridges.

(ii) *Tacoma Bridge* (Washington, 1940)

On July 1, 1940, the Tacoma Narrows Bridge was opened to traffic. From this first day, the bridge began undergoing vertical oscillations, and was nicknamed 'Galloping Gertie'. For the next four months people came from far and wide to enjoy the thrill of riding over the galloping bridge. But on November 7, early in the morning, the bridge began undulating persistently for three hours, and at 10.30 a.m. the bridge began cracking, and finally at 11.10 a.m. the entire bridge came crashing down, the only life lost being that of a reporter's pet dog which was abandoned along with the reporter's car on the bridge.

5.3 CONSUMER BUYING BEHAVIOUR

In §4.2 we described a mathematical model which attempted to estimate the effect of advertising on the sales of a particular article. This section considers the decision process of an individual consumer, and so formulates a model for predicting consumer behaviour towards a particular product, say brand X.

The basic variables, with t time, are

$B(t)$ – buying level for brand X
$M(t)$ – motivation (or attitude) towards brand X
$C(t)$ – level of communication (e.g. advertising) of brand X

and these variables are assumed to be related by

$$\frac{\mathrm{d}B}{\mathrm{d}t} = b(M - \beta B) \qquad (5.22)$$

$$\frac{\mathrm{d}M}{\mathrm{d}t} = a(B - \alpha M) + \gamma C \qquad (5.23)$$

where a, b, α, β and γ are constants which for most articles are positive.

Although we have a system of two first order coupled differential equations, we can substitute M from (5.22) into (5.23) to give

$$\frac{\mathrm{d}}{\mathrm{d}t}\left(\frac{1}{b}\frac{\mathrm{d}B}{\mathrm{d}t} + \beta B\right) = a\left(B - \frac{\alpha}{b}\frac{\mathrm{d}B}{\mathrm{d}t} - \alpha\beta B\right) + \gamma C$$

i.e.
$$\boxed{\frac{d^2B}{dt^2} + (b\beta + a\alpha)\frac{dB}{dt} + ab(\alpha\beta - 1)B = b\gamma C}. \qquad (5.24)$$

This is again a second order linear differential equation which relates the buying level of brand X to the communication level C.

From §5.1 we know that the solution of this equation is of the form

$$B = B_c + B_p$$

where B_p is a particular solution, which will depend on the communication level C, and B_c is the general solution of the associated homogeneous equation

$$\boxed{\frac{d^2B}{dt^2} + (b\beta + a\alpha)\frac{dB}{dt} + ab(\alpha\beta - 1)B = 0}. \qquad (5.25)$$

We first determine B_c. From §5.1 the auxiliary equation is

$$m^2 + (b\beta + a\alpha)m + ab(\alpha\beta - 1) = 0$$

giving

$$m = \{-(b\beta + a\alpha) \pm [(b\beta + a\alpha)^2 - 4ab(\alpha\beta - 1)]^{1/2}\}/2$$

$$= \{-(b\beta + a\alpha) \pm [(b\beta - a\alpha)^2 + 4ab]^{1/2}\}/2$$

So $m = m_1$ and m_2, where

$$m_1 = \{-(b\beta + a\alpha) + [(b\beta - a\alpha)^2 + 4ab]^{1/2}\}/2$$

$$m_2 = \{-(b\beta + a\alpha) - [(b\beta - a\alpha)^2 + 4ab]^{1/2}\}/2.$$

Clearly if all the parameters are positive, then m_1 and m_2 will be real, and $m_2 < 0$. The sign of m_1 depends on the sign of $(\alpha\beta - 1)$. If $\alpha\beta > 1$, then it can be shown that $m_1 < 0$, whereas if $\alpha\beta < 1$, $m_1 > 0$. So we can write the solution as

$$B_c = Ae^{m_1 t} + Be^{m_2 t} \qquad (5.26)$$

and we note that $B_c \to 0$ provided $\alpha\beta < 1$.

Now we turn to B_p, which obviously depends on the form of C. Taking the simplest case to analyse, suppose the communication level is kept constant, i.e.

$$C = \bar{C} \text{ for all time } t > 0. \tag{5.27}$$

Then it is easy to see that a particular solution of

$$\frac{d^2B}{dt^2} + (b\beta + a\alpha)\frac{dB}{dt} + ab(\alpha\beta - 1)B = b\gamma\bar{C}$$

is given by $B_p = \gamma\bar{C}/a(\alpha\beta - 1)$. There are two cases to consider

(i) $\alpha\beta > 1$ — Then from (5.26), we know that m_1, $m_2 < 0$ and so $B_c \to 0$ as $t \to \infty$, and

$$B \to B_p = \gamma\bar{C}/a(\alpha\beta - 1) \tag{5.28}$$

In other words, the buying level tends to an equilibrium level, which of course depends on the magnitude of $\bar{C}$. This is illustrated in Fig. 5.8.

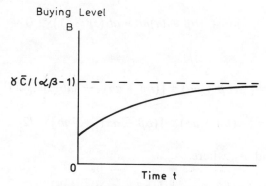

Fig. 5.8—Buying Level for $\alpha\beta > 1$

(ii) $\alpha\beta < 1$ — Now the particular solution is negative, but this is of no significance, because the other part of the solution, B_c, no longer tends to zero as $t \to \infty$. In fact, as illustrated in Fig. 5.9, B_c, and so B, tends to ∞ as $t \to \infty$, and we have an unstable situation.

For practical application, we would expect $\alpha\beta > 1$, and the behaviour illustrated in case (i) to be more likely to occur.

5.4 ELECTRICAL NETWORKS

In this section we extend the analysis started in §4.4, applying the theory to a slightly more complicated network as shown in Fig. 5.10.

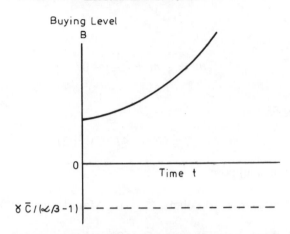

Fig. 5.9—Buying Level for $\alpha\beta < 1$

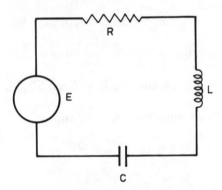

Fig. 5.10—Electrical Network

This circuit has a resistor, an inductor and a capacitor in series with an electromotive force. Applying Kirchhoff's Law, we obtain

$$L\frac{\mathrm{d}I}{\mathrm{d}t} + RI + \frac{Q}{C} = E, \tag{5.29}$$

and since $\mathrm{d}Q/\mathrm{d}t = I$, we can differentiate (5.29) to obtain

$$\boxed{L\frac{\mathrm{d}^2I}{\mathrm{d}t^2} + R\frac{\mathrm{d}I}{\mathrm{d}t} + \frac{I}{C} = 0} \tag{5.30}$$

when E is a constant.

This is a second order linear homogeneous equation with auxiliary equation

$$m^2 + \frac{R}{L}m + \frac{1}{CL} = 0, \qquad (5.31)$$

which has roots

$$m = [-R \pm (R^2 - 4L/C)^{1/2}]/2L. \qquad (5.32)$$

For example, suppose

$L = 1$ henry, $R = 100$ ohms, $C = 10^{-4}$ farads, $E = 1000$ volts,

and at $t = 0$, there are no charges ($Q = 0$) and no current flowing ($I = 0$) when E is applied; then from (5.32)

$$m = -50 \pm 50\sqrt{3}i$$

giving, from (5.7), solution

$$I = e^{-50t}(A \cos 50\sqrt{3}t + B \sin 50\sqrt{3}t).$$

Since $I = 0$ at $t = 0$, we must have $A = 0$, and so

$$I = Be^{-50t} \sin 50\sqrt{3}t.$$

Now, from (5.29),

$$Q = C[E - L \, dI/dt - RI]$$

$$= 1/10 - Be^{-50t}[\sin 50\sqrt{3}t + \sqrt{3} \cos 50\sqrt{3}t]/200$$

and since $Q = 0$ at $t = 0$, we have $B = 20/\sqrt{3}$. Thus

$$I = 20e^{-50t} \sin 50\sqrt{3}t/\sqrt{3} \qquad (5.33)$$

$$Q = 1/10 - e^{-50t}[\sin 50\sqrt{3}t + \sqrt{3} \cos 50\sqrt{3}t]/10\sqrt{3}. \qquad (5.34)$$

These equations show that the current soon decays to zero whilst the charge tends to its steady state value 1/10 coulombs, as illustrated in Fig. 5.11.

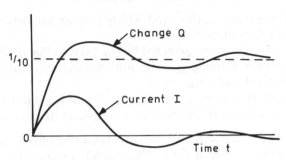

Fig. 5.11—Current and 'Charge

5.5 TESTING FOR DIABETES

Diabetes mellitus is a disease of metabolism which is characterised by too much sugar in the blood and urine. In diabetes the body is unable to burn off all its sugar, starches and carbohydrates because of an insufficient supply of insulin. Diabetes is usually diagnosed by means of a glucose tolerance test (GTT). The patient comes to hospital after an overnight fast, is given a large dose of glucose, and during thc following hours measurements are regularly made of the concentration of glucose in the bloodstream. In the mid 1960's Drs Rosevear and Molnar of the Mayo Clinic, and Ackerman and Gatewood of the University of Minnesota discovered a fairly reliable criterion for interpreting GTT results.

They developed a model for the blood glucose regulatory system, which is based on the following assumptions:

1. *Glucose* is a *source of energy* for all tissues and organs, and for each individual there is an optimal blood glucose concentration. Excessive deviations from this optimal value cause severe problems.
2. Blood glucose levels are influenced by a wide variety of hormones and other metabolites such as
   (i) *Insulin*, which is secreted by the β cells of the pancreas. After eating carbohydrates, the pancreas secretes more insulin. In addition, the glucose in our blood directly stimulates the β cells to secrete insulin. The insulin facilitates the tissue uptake of glucose.
   (ii) *Glucagon*, a hormone secreted by the α cells of the pancreas. Excess glucose is stored in the liver in the form of glucogen, and in times of need it is converted back into glucose.
   (iii) *Epinephrine* (adrenaline), a hormone secreted by the adrenal medulla, which is part of an emergency mechanism to increase the concentration of glucose in the blood in times of extreme hypoglycemia (low blood sugar).
   (iv) *Glucocorticoids*, hormones such as cortisone which are secreted

by the adrenal cortex, and which play an important role in the metabolism of carbohydrates.

(v) *Thyroxin*, hormone secreted by the thyroid gland, and which aids the liver in forming glucose from non-carbohydrate sources such as glycerol and amino acids.

(vi) *Growth Hormone*, secreted by the anterior pituitary gland, and which tends to block insulin, i.e. reducing its effectiveness.

The model postulated is a simple one, requiring only a limited number of blood samples during a GTT. The models centre on two concentrations, namely

(i) G, glucose in the blood
(ii) H, net hormonal concentration

The latter represents the cumulative effect of all the important hormones; insulin is considered to increase H whilst cortisone decreases H. The basic model is then described by the equations

$$\frac{dG}{dt} = F_1(G, H) + J(t) \tag{5.35}$$

$$\frac{dH}{dt} = F_2(G, H) \tag{5.36}$$

where $J(t)$ is the external rate at which the blood glucose concentration is being increased.

We assume that the G and H have achieved equilibrium values, say G_0 and H_0, by the time the fasting patient arrives at the hospital, i.e.

$$F_1(G_0, H_0) = 0 = F_2(G_0, H_0)$$

Now we are interested in the deviations of G, H from their equilibrium values, so we take

$$G = G_0 + g$$

$$H = H_0 + h$$

where g and h are small compared to G_0 and H_0. The governing equations become

$$\frac{dg}{dt} = F_1(G_0 + g, H_0 + h) + J(t)$$

$$= f_1(g, h) + J(t), \quad \text{say}$$

and

$$\frac{dh}{dt} = f_2(g, h), \quad \text{say;}$$

and we assume the function f_1 and f_2 have a linear form. Thus

$$\frac{dg}{dt} = -ag - bh + J(t) \tag{5.37}$$

$$\frac{dh}{dt} = -ch + eg. \tag{5.38}$$

Now $a > 0$, since $dg/dt < 0$ for $h = 0$ through tissue uptake of glucose, and $b > 0$, since $h > 0$ tends to decrease blood glucose levels. Also $c > 0$ since the concentration of hormones in the blood decrease through hormone metabolism and $e > 0$ for $g > 0$ causes the endocrine glands to secrete those hormones which tend to increase h.

From (5.37),

$$h = (-ag + J - dg/dt)/b$$

and substituting into (5.38) gives

$$-a\frac{dg}{dt} + \frac{dJ}{dt} - \frac{d^2g}{dt^2} = -c\left(-ag + J - \frac{dg}{dt}\right) + beg$$

i.e.
$$\frac{d^2g}{dt^2} + (a + c)\frac{dg}{dt} + (ac + be)g = \frac{dJ}{dt} + cJ. \tag{5.39}$$

Now the right hand side is zero except over a small time interval. Let time $t = 0$ be the time the glucose load has been ingested, so that for $t > 0$,

$$\boxed{\frac{d^2g}{dt^2} + (a + c)\frac{dg}{dt} + (ac + be)g = 0} \ . \tag{5.40}$$

We can rewrite this as

$$\frac{d^2g}{dt^2} + 2\alpha\frac{dg}{dt} + \omega^2 g = 0$$

which has auxiliary equation

$$m^2 + 2\alpha m + \omega^2 = 0.$$

If $\alpha^2 - \omega^2 < 0$, the solution is given by

$$g = e^{-\alpha t}(A \cos(\omega_0 t) + B \sin(\omega_0 t))$$

where $\omega_0 = \omega - \alpha$. Thus the complete solution is

$$G = G_0 + e^{-\alpha t}(A \cos(\omega_0 t) + B \sin(\omega_0 t)) \tag{5.41}$$

and this contains *five* unknowns

$$G_0, \alpha, A, \omega_0, B$$

We can of course determine G_0 initially before the glucose is taken. Then if we take four measurements G_i at time t_i ($i = 1, 2, 3, 4$), we have four equations for four unknowns α, A, ω_0, B. Better than that, we take, say, n measurements, and use 'least squares' techniques to find the best fit for the parameters—this has to be solved on a computer.

Observations have shown that slight errors in measuring G can produce large errors in the value of α. However, the value of the parameter ω_0, was relatively insensitive to experimental errors in G. So we determine ω_0, and use this value as a basic description for the response to a GTT. The remarkable fact is that a value for $T_0 = 2\pi/\omega_0$ of *less than 4 hours* indicated *normality*, whilst appreciably *more than 4 hours* indicated *mild diabetes*.

EXERCISES

1. **Mechanical Oscillations**
   (i) Using the model of mechanical oscillations described in §5.2 determine the predicted motion when resistance is ignored ($k = 0$), and

   (a) $F = 0$ for all $t > 0$, and $x = a$, $dx/dt = 0$ at $t = 0$
   (b) $F = F_0 \cos \alpha t$, and $x = 0$, $dx/dt = 0$ at $t = 0$

   Is the solution to (b) valid for all α? Determine the solution when $\alpha = \omega$, and illustrate it graphically.

   (ii) With the model described in §5.2 including resistance, determine the predicted motion when

   (a) $F = F_0$, constant for all $t > 0$;
   (b) $F = F_0 e^{-t}$ for all $t > 0$.

2. **Consumer Buying Behaviour**

    With the model described in §5.3, determine the predicted buying level when

    (i) $c = \begin{cases} \bar{c}, \text{ constant} & 0 \leq t \leq T \\ 0 & t > T \end{cases}$

    (ii) $c = \bar{c}e^{-t}$ for $t \geq 0$.

3. **Electrical Networks**

    For the electrical network shown in Fig. 5.10, determine the current and charge when

    $L = 1$ henry, $R = 100$ ohms, $C = 10^{-4}$ farads, $E = 1000 \cos 100t$ volts

    and initially there is no charge and no current flowing.

4. **National Economy Model**

    A simplified model for a national economy is described in terms of the following variables

    Y – national output
    I – investment
    G – government spending
    C – consumption

    The demand, D, for goods and services is given by

    $$D = C + I + G$$

    and it is assumed that $C = (1 - s)Y$, where s is the savings coefficient. We also assume that ouput responds to the excess demand so that

    $$\frac{dY}{dt} = l(D - Y)$$

    where l is a constant; and investment responds according to

    $$\frac{dI}{dt} = m\left(a\,\frac{dY}{dt} - I\right)$$

    where a and m are constants. If G is taken as a constant, show that Y

satisfies an equation of the form

$$\frac{d^2Y}{dt^2} + \alpha \frac{dY}{dt} + \beta Y = mlG.$$

What predictions can you make about the type of behaviour indicated by this model?

5. **Pricing Policy**

A general pricing policy for a manufacturer is described by the differential equation

$$\frac{dp}{dt} = -\delta(L(t) - L_0)$$

where δ is a positive constant, $L = L(t)$ is the inventory level at time t and L_0 the desired inventory level. The inventory changes according to

$$\frac{dL}{dt} = Q - S$$

where Q and S are the production and sales respectively, and these are modelled by

$$Q = a - bp - c\frac{dp}{dt}$$

$$S = \alpha - \beta p - \gamma \frac{dp}{dt}$$

where a, b, c, α, β and γ are positive constants. Show that the forecast price is determined from

$$\frac{d^2p}{dt^2} + \delta(\gamma - c)\frac{dp}{dt} + \delta(\beta - b) = \delta(\alpha - a).$$

If $\alpha > a$, $\beta > b$, show that the price is tending to a stable value as t increases.

6. **Competing Species**

A model for two species competing for their food supply is given by

$$\frac{dx}{dt} = ax - by$$

$$\frac{dy}{dt} = cy - dx$$

where x, y are the two species populations and a, b, c and d are positive constants. Show that x satisfies

$$\frac{d^2x}{dt^2} - (a + c)\frac{dx}{dt} + (ac - bd)x = 0$$

and deduce that x has solution of the form

$$x = Ae^{\alpha_1 t} + Be^{\alpha_2 t}$$

where at least one of the α_i is positive. Find also the solution for y.
   Using the parameter values

$$a = c = 2, \qquad b = d = 1$$

and with $x = 100$, $y = 200$ at time $t = 0$, determine the time when one species is eliminated.

Chapter 6

Non-Linear Second Order Differential Equations

6.1 INTRODUCTION

In the preceding chapter we have seen how to construct the solutions of *linear* second order differential equations. Unfortunately no such methods are available for *non-linear* second order differential equations, and in general, most cannot be solved analytically.

We can write second order differential equations in general as

$$\boxed{\frac{d^2y}{dx^2} = f\left(y, \frac{dy}{dx}, x\right)} \tag{6.1}$$

where f is an arbitrary function of y, dy/dx and x. There are a number of special types of (6.1) which can be reduced to first order differential equations. For example, if y is missing in f, then we can use a substitution $p = dy/dx$ to reduce (6.1) to a first order differential equation,

$$\frac{dp}{dt} = f(p, x).$$

Example
Solve the differential equation

$$\frac{dy}{dx} \cdot \frac{d^2y}{dx^2} = 1.$$

Solution
Using the substitution $p = dy/dx$, then $d^2y/dx^2 = dp/dx$ and

$$p \frac{dp}{dx} = 1$$

Separating the variables of the first order differential equation gives

$$\int p \, dp = \int dx$$

i.e. $\frac{1}{2}p^2 = x + A$ (A arbitrary constant).

Thus

$$p = \pm(2x + 2A)^{1/2}$$

i.e. $\frac{dy}{dx} = \pm(2x + 2A)^{1/2},$

and integrating again

$$y = \pm(1/3)(2x + 2A)^{3/2} + B \quad (B \text{ arbitrary constant}).$$

6.2 PLANETARY MOTION

Suppose a planet of mass M is moving round the sun on a path with position vector $\mathbf{r} = \mathbf{r}(t)$, and assuming Newton's inverse square gravitational law, the equation of motion for the planet is

$$M \frac{d^2\mathbf{r}}{dt^2} = -\frac{\gamma M_s M}{r^2} \mathbf{e}_r. \tag{6.2}$$

Here γ is the universal gravitational constant, M_s is the mass of the sun and $\mathbf{e}_r$ is a unit vector in the direction of $\mathbf{r}$. Writing $k = \gamma M_s$ and assuming motion in a plane, with plane polar coordinates (r, θ), (6.2) becomes

$$\frac{d^2\mathbf{r}}{dt^2} = -\frac{k}{r^2} \mathbf{e}_r, \tag{6.3}$$

and $\mathbf{r} = r\mathbf{e}_r$, as illustrated in Fig. 6.1.

Now

$$\frac{d\mathbf{r}}{dt} = \dot{r}\mathbf{e}_r + r\dot{\theta}\mathbf{e}_\theta$$

$$\frac{d^2\mathbf{r}}{dt^2} = (\ddot{r} - r\dot{\theta}^2)\mathbf{e}_r + (2\dot{r}\dot{\theta} + r\ddot{\theta})\mathbf{e}_\theta,$$

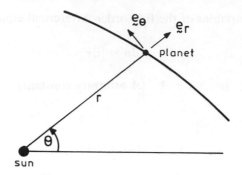

Fig. 6.1—Planetary Motion about Sun

where 'dot' denotes time differentiation. Substituting for d^2r/dt^2 in (6.3) gives

$$(\ddot{r} - r\dot{\theta}^2)\mathbf{e}_r + (2\dot{r}\dot{\theta} + r\ddot{\theta})\mathbf{e}_\theta = -(k/r^2)\mathbf{e}_r, \qquad (6.4)$$

and equating coefficients of $\mathbf{e}_r$ and $\mathbf{e}_\theta$ gives

$$\ddot{r} - r\dot{\theta}^2 = -k/r_2 \qquad (6.5)$$

$$2\dot{v}\dot{\theta} + v\ddot{\theta} = 0. \qquad (6.6)$$

Now the second equation (6.6) can be written as

$$\frac{1}{r}\frac{d}{dt}(r^2\dot{\theta}) = 0.$$

Thus $\dfrac{d}{dt}(r^2\dot{\theta}) = 0$, and integrating

$$r^2\dot{\theta} = h, \text{ constant} \qquad (6.7)$$

(h is in fact the angular momentum per unit mass about the sun). Substituting for $\dot{\theta}$ in (6.5) now gives

$$\boxed{\ddot{r} - h^2/r^3 = -k/r^2} \quad , \qquad (6.8)$$

which is a non-linear second order differential equation for r as a function of t.

We can solve this equation by using the substitution $p = 1/r$ and solving for p as a function of θ rather than t. Now

$$\frac{dp}{d\theta} = \frac{dp}{dt}\frac{dt}{d\theta} = \frac{d}{dt}\left(\frac{1}{r}\right) / \dot\theta = -\dot r/r^2\dot\theta = -\dot r/h,$$

using (6.7). Also

$$\frac{d^2p}{d\theta^2} = -\frac{1}{h}\frac{d}{d\theta}(\dot r) = -\frac{1}{h}\frac{d}{dt}(\dot r)\frac{dt}{d\theta} = -\ddot r/h\dot\theta = -r^2\ddot r/h^2$$

again using (6.7). Thus (6.8) becomes

$$-h^2p^2\frac{d^2p}{d\theta^2} - p^3h^2 = -p^2k$$

or

$$\boxed{\frac{d^2p}{d\theta^2} + p = k/h^2} \ . \tag{6.9}$$

This is a linear second differential equation, and using the methods of §5.1, the complementary solution is $p = A\cos\theta + B\sin\theta$, and one particular solution is $p = k/h^2$. Choosing axis so that $B = 0$, we can write the solution as

$$p = k/h^2 + A\cos\theta, \tag{6.10}$$

and remembering $p = 1/r$ gives

$$\boxed{r = \frac{(h^2/k)}{1 + (Ah^2/k)\cos\theta}} \ . \tag{6.11}$$

What does this predict?

Using polar coordinates, the equation of a conic is

$$r = l/(1 + e\cos\theta)$$

where l is called the semi-latus rectum and e the eccentricity ($e > 1$ is a hyperbola, $e = 1$ is a parabola, $0 \leqslant e < 1$ is an ellipse with $e = 0$ a circle). Now the only closed conic is an ellipse, and since planets move on closed orbit, we can conclude that (6.11) predicts that planets move on *elliptic* orbits, with eccentricity $e = Ah^2/k$ and semi-latus rectum $l = h^2/k$.

Solar system data is given in Table 6.1

Table 6.1—Solar System Data

Planetary Body	Mean Distance from SUN (compared with the Earth)	Mass (compared with the Earth)	Volume (compared with the Earth)	Radius (compared with the Earth)	Period of Revolution about Sun (Years)	Period of Rotation about own axis (Days)	Eccentricity of Orbit	Number of Moons
Sun	—	333,434	1,301,813	109.187(?)	—	24.65	—	—
Mercury	0.387	0.04	0.055	0.366	0.24	0.59	0.206	0
Venus	0.723	0.83	0.904	0.960	0.61	30.0(?)	0.007	0
Earth	1.000	1.00	1.000	1.000	1.00	1.00	0.017	1
Mars	1.524	0.11	0.150	0.273	1.90	1.03	0.093	2
Jupiter	5.204	318.0	1318.7	10.969	11.9	0.41	0.048	12
Saturn	9.5	95.0	767.2	9.036	29.5	0.42	0.056	9
Uranus	19.2	15.0	49.4	3.715	84.0	0.45	0.047	5
Neptune	30.1	17.0	41.8	3.538	165.0	0.66	0.009	2
Pluto	39.5	0.8(?)	1.07(?)	1.02(?)	248.0	?	0.249	0
Earth's Moon	—	0.012	0.020	0.273	—	27.32	—	—
	Earth = 149.5×10^6 km	Earth = 5.976×10^{24} kg	Earth = 1.08×10^{12} km^3	Earth = 6368 km				

Gravitational constant

$$\gamma = 6.67 \times 10^{-11} \text{ m}^3 \text{ kg}^{-1} \text{ s}^2$$

For most planets the eccentricity of their elliptic orbit is small, and as a first approximation we can assume a circular orbit (i.e. take $A = 0$ in (6.11)) $r = h^2/k$. Now from (6.7)

$$\frac{d\theta}{dt} = \frac{h}{r^2} = \frac{k^{1/2}}{r^{3/2}},$$

and integrating round a complete orbit gives

$$2\pi = (k^{1/2}/r^{3/2})T$$

where T is the planet's periodic time. So the model predicts that

$$\boxed{T = (2\pi/k^{1/2})r^{3/2}} \qquad (6.12)$$

i.e. $T \alpha r^{3/2}$. We can check this prediction by noting that (6.12) can be written, after taking logs, as

$$\ln T = K + 3/2 \ln r$$

where K is a constant, and so a log $T - \ln r$ graph should be a straight line with slope $3/2$. Using the data from Table 6.1 we obtain the graph, Fig. 6.2, which shows excellent correlation between the data and model.

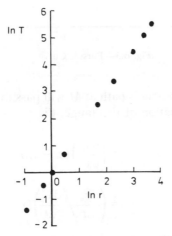

Fig. 6.2—Keplar's Third Law, $\tau \alpha r^{3/2}$

(This is Kepler's third law of planetary motion, which he deduced from the observed data.)

6.3 PURSUIT CURVES

Guided missiles have various mechanisms for moving towards their target. For example, some are heat-seeking and always move towards the exhaust from an aeroplane. Once the guidance system of the missile has locked onto the aeroplane, the missile will move so that it is always pointing in the direction of the aeroplane. We will show how to find the missile's path and also find the capture time.

We introduce coordinate axes such that at $t = 0$, the missile is at the origin $(0, 0)$ and the aeroplane at (a, b). As illustrated in Fig. 6.3, we suppose the aeroplane moves parallel to the x-axis with constant speed v_A. We also suppose that the missile has constant speed v_m, and we denote its coordinates by (x_m, y_m).

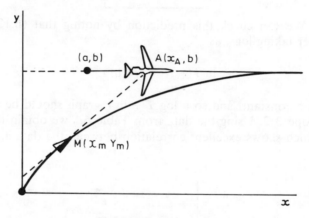

Fig. 6.3—Pursuit Curve

The tangent to the missile's path at M will pass through the position of the aeroplane. The equation of this tangent is

$$y - y_m = \left(\frac{dy_m}{dx_m}\right)(x - x_m)$$

$$= \left(\frac{dy_m}{dt} \middle/ \frac{dx_m}{dt}\right)(x - x_m).$$

Now $A(x_A, b)$ lies on this line, and $x_A = a + v_A t$, giving

$$b - y_m = (\dot{y}_m / \dot{x}_m)(a + v_A t - x_m)$$

and where

$$v_m^2 = \dot{x}_m^2 + \dot{y}_m^2.$$

Dropping the suffix m, the path of the missile is given by the solution of

$$\dot{x}(b - y) = \dot{y}(a + v_A t - x) \tag{6.13}$$

$$\dot{x}^2 + \dot{y}^2 = v_m^2 \tag{6.14}$$

with boundary conditions $x(0) = y(0) = 0$. This is a system of two non-linear differential equations. By writing (6.13) as

$$\frac{dx}{dy}(b - y) = a + v_A t - x$$

and differentiating with respect to t, we obtain

$$\frac{d^2x}{dy^2}\frac{dy}{dt}(b - y) - \frac{dx}{dy}\frac{dy}{dt} = v_A - \frac{dx}{dt},$$

i.e. $$\frac{d^2x}{dy^2}\frac{dy}{dt}(b - y) = v_A. \tag{6.15}$$

From (6.14),

$$\dot{y}^2(1 + (\dot{x}/\dot{y})^2) = v_m^2 ,$$

i.e. $$\frac{dy}{dt} = v_m/[1 + (dx/dy)^2]^{1/2};$$

and substituting in (6.15) gives

$$\boxed{\frac{d^2x}{dy^2}(b - y) = c[1 + (dx/dy)^2]^{1/2}} , \quad c = v_A/v_m. \tag{6.16}$$

which is a second order non-linear differential equation.

We can solve this by using the substitution $p = dx/dy$, so that $dp/dy = d^2x/dy^2$ and (6.16) becomes

$$\frac{dp}{dy}(b - y) = c(1 + p^2)^{1/2}.$$

This is a variables separable differential equation (of the type met in Chapter 3) which we can integrate to give

$$\int \frac{dp}{(1 + p^2)^{1/2}} = c \int \frac{dy}{(b - y)}$$

i.e. $\qquad \ln[p + (1 + p^2)^{1/2}] = -c \ln(b - y) + K.$

Now, initially, $y = 0$ and $p (= dx/dy) = a/b = d$, say. Thus

$$K = \ln[d + (1 + d^2)^{1/2}] + c \ln b = \ln(fb^c),$$

where $f = d + (1 + d^2)^{1/2}$. Hence

$$\ln[p + (1 + p^2)^{1/2}] = \ln(b - y)^{-c} + \ln(fb^c) = \ln[fb^c/(b - y)^c]$$

and

$$p + (1 + p^2)^{1/2} = fb^c/(b - y)^c. \qquad (6.17)$$

We can obtain p from (6.17) by writing

$$(1 + p^2)^{1/2} = [fb^c/(b - y)^c] - p$$

and squaring,

$$1 + p^2 = f^2 b^{2c}/(b - y)^{2c} - 2pfb^c/(b - y)^c + p^2,$$

and rearranging,

$$\frac{dx}{dy} = \frac{1}{2}\left\{\frac{fb^c}{(b - y)^c} - \frac{(b - y)^c}{fb^c}\right\}. \qquad (6.18)$$

Integrating with respect to y,

$$x = \frac{1}{2}\left\{\frac{fb^c}{(c - 1)(b - y)^{c-1}} + \frac{(b - y)^{c+1}}{(c + 1)fb^c}\right\} + K_1$$

(and $c \neq 1$, since we assume $c < 1$, i.e. $v_A < v_m$). Now K_1 is evaluated from the initial conditions $x = y = 0$, i.e.

$$0 = 1/2\{fb/(c - 1) + b/(c + 1)f\} + K_1.$$

Thus

$$K_1 = b[(f^2 + 1)c + f^2 - 1]/2f(1 - c^2) \qquad (6.19)$$

and so the missile's path is given by

$$x = \frac{1}{2}\left\{ \frac{(b - y)^{c+1}}{(c + 1)fb^c} - \frac{fb^c(b - y)^{1-c}}{(1 - c)} + K_1 \right\}. \qquad (6.20)$$

The missile hits its target when its coordinates are the same as that of the aeroplane, i.e. $x = a + v_A t$, $y = b$; and substituting these values in (6.20) gives

$$a + v_A t = K_1,$$

which after some algebra gives the capture time as

$$t = \frac{[(a^2 + b^2)^{1/2} + ac]}{v_m(1 - c^2)}, \quad c = v_A/v_m. \qquad (6.21)$$

The path for the missile using parameter values

$$a = 5000 \text{ m}, b = 3000 \text{ m}, v_A = 1000 \text{ ms}^{-1}, v_m = 2000 \text{ ms}^{-1}$$

is shown in Fig. 6.4. The capture time is given from (6.21) as 5.55 s, and the capture position is (10,554, 3000).

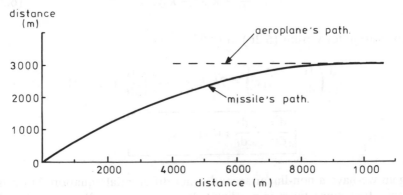

Fig. 6.4—Pursuit Curve Solution

6.4 CHEMICAL KINETICS

Chemical reactions are governed by the Law of Mass Action, which states that the rate of a reaction is proportional to the active concentrations of the reactants. For example, if a molecule each of A and B combine reversibly to form C we write

$$A + B \underset{k_2}{\overset{k_1}{\rightleftharpoons}} C, \tag{6.22}$$

and if x_1, x_2, x_3 are the concentrations of A, B, C respectively, the law of mass action gives

$$\frac{dx_1}{dt} = \frac{dx_2}{dt} = k_2 x_3 - k_1 x_1 x_2 \tag{6.23}$$

$$\frac{dx_3}{dt} = k_1 x_1 x_2 - k_2 x_3, \tag{6.24}$$

where k_1, k_2 are the reaction rates.

We will study an even simpler reaction, described by

$$A + A \underset{k_2}{\overset{k_1}{\rightleftharpoons}} A_2. \tag{6.25}$$

If x is the concentration of A, y of A_2, then

$$\frac{dx}{dt} = 2k_2 y - 2k_1 x^2 \tag{6.26}$$

$$\frac{dy}{dt} = k_1 x^2 - k_2 y. \tag{6.27}$$

Substituting for y from (6.26) in (6.27) gives

$$\frac{d}{dt}\left(\frac{dx}{dt} + 2k_1 x^2\right) = 2k_2 k_1 x^2 - k_2\left(\frac{dx}{dt} + 2k_1 x^2\right),$$

i.e.

$$\boxed{\frac{d^2 x}{dt^2} + \frac{dx}{dt}(4k_1 x + k_2) = 0}. \tag{6.28}$$

Again we have a non-linear second order differential equation, but it is a special type since there is no direct dependence on t. We can solve this type of equation by using the substitution $p = dx/dt$, but solving for p as a function of x, rather than t. We have

$$\frac{d^2x}{dt^2} = \frac{dp}{dt} = \frac{dp}{dx}\frac{dx}{dt} = p\frac{dp}{dx},$$

so that (6.28) becomes

$$p\frac{dp}{dx} + p(4k_1x + k_2) = 0. \tag{6.29}$$

Assuming $p \neq 0$ (if $p = 0$, $dx/dt = 0$ and we have equilibrium),

$$\frac{dp}{dx} = -4k_1x - k_2,$$

and, integrating,

$$p = \frac{dx}{dt} = -2k_1x^2 - k_2x + c_0. \tag{6.30}$$

We now have a first order differential equation and we can separate the variables to give

$$\int \frac{dx}{2k_1x^2 + k_2x - c_0} = -\int dt,$$

i.e.

$$\frac{1}{2k_1}\int \frac{dx}{(x - \alpha_1)(x - \alpha_2)} = -t + c_1,$$

where α_1, α_2 are the roots of $x^2 + (k_2/2k_1)x - (c_0/2k_1) = 0$. This gives

$$\alpha_1 = (-k_2 + a)/4k_1, \quad \alpha_2 = -(k_2 + a)/4k_1,$$

where $a = (k_2^2 + 8c_0k_1)^{1/2}$.

Partial fractioning the integrand gives

$$\frac{1}{2k_1(\alpha_1 - \alpha_2)}\int \left[\frac{1}{(x - \alpha_1)} - \frac{1}{(x - \alpha_2)}\right] dx = -t + c_1$$

with $\alpha_1 - \alpha_2 = a/2k_1$. Integrating

$$\frac{1}{a}\log\left(\frac{x - \alpha_1}{x - \alpha_2}\right) = -t + c_1,$$

and rearranging

$$x = \frac{(k_2 + a)c_2 e^{-at} + a - k_2}{4k_1(1 - c_2 e^{-at})}, \qquad (6.31)$$

where $c_2 = e^{ac_1}$. Substituting (6.30) into (6.26) gives

$$2k_2 y = -k_2 x + c_0; \qquad (6.32)$$

so if the initial concentrations x_0, y_0 are given, then

$$c_0 = 2k_2 y_0 + k_2 x_0. \qquad (6.33)$$

Also, from (6.31),

$$c_2 = \frac{4k_1 x_0 - a + k_2}{4k_1 x_0 + a + k_2}. \qquad (6.34)$$

We could now obtain the complete solutions for x and y, but more important is the approximate behaviour as t gets large. From (6.31), we see that as $t \to 0$,

$$x \to \frac{a - k_2}{4k_1} = \frac{(k_2^2 + 8c_0 k_1)^{1/2} - k_2}{4k_1} = x^*, \quad \text{say,}$$

and from (6.32),

$$y \to \frac{k_2 - (k_2^2 + 8c_0 k_1)^{1/2}}{8k_1} + \frac{c_0}{2k_2} = y^*, \quad \text{say.}$$

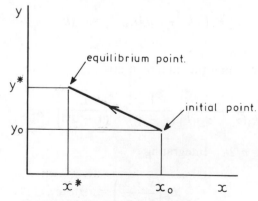

Fig. 6.5—Phase plane for Chemical Reaction

So as t increases, we tend to an equilibrium solution. This is illustrated in an x, y graph, Fig. 6.5.

EXERCISES

1. Inverse Cube Law

Show that motion of a planet about a sun, assuming an inverse *cube* gravitational law, can be reduced to

$$\frac{d^2r}{dt^2} - \frac{h^2}{r^3} = -\frac{\lambda}{r^3},$$

λ constant. Use the substitution $p = 1/r$, and deduce that

$$\frac{d^2p}{d\theta^2} + (1 - \lambda/h^2)p = 0.$$

Determine the possible orbits.

2. Pursuit Curves

(i)   The y-axis and the line $x = a$ are banks of a river, as shown in Fig. 6.6. The river flows in the negative y direction with speed u.

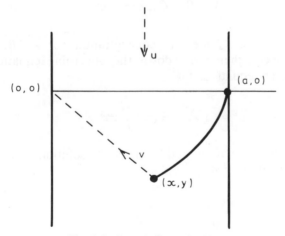

Fig. 6.6—Pursuit Curve for boat

A boat whose speed in still water is v is launched from $(a, 0)$. The boat is steered so that it is always headed towards the origin. Find the path of the boat. Under what conditions will the boat land on the opposite bank? Where will it land?

(ii) Suppose that a hawk P at the point $(a, 0)$ spots a pigeon Q at the origin of coordinates flying along the y-axis at a speed v. The hawk immediately flies towards the pigeon at a speed w. What will be the flight path of the hawk? Distinguish between the cases $v < w, v = w, v > w$.

(iii) A destroyer is in dense fog, which lifts for an instant disclosing an enemy submarine on the surface 4 km away. The submarine dives immediately and proceeds at full speed in an unknown direction. What path should the destroyer select to be certain of passing over the submarine, if its velocity v is three times that of the submarine?

[Hint: There is not a unique answer to this problem. One method is for the destroyer to travel 3 km towards the point where the submarine was spotted, and then move on a path $v = f(\theta)$.]

3. **Chemical Kinetics**

(i) For the bimolecular reaction described by equations (6.26) and (6.27) find the solution if the concentrations of A and B are the same.

(ii) A chemical reaction is described by

$$A + B + C \underset{k_2}{\overset{k_1}{\rightleftharpoons}} D.$$

If x_1, x_2, x_3, x_4 are the concentrations of A, B, C and D respectively, then write down the governing equations. Hence deduce the equations for

$$A + A + A \underset{k_2}{\overset{k_1}{\rightleftharpoons}} A_3.$$

Solve these equations and illustrate the solutions in a graph of the concentration of A against A_3.

Chapter 7

Systems of Differential Equations

7.1 INTRODUCTION

We have already met systems of differential equations in sections 5.3 and 5.5, but in both cases the systems of two linear first order differential equations were reduced to single second order linear equations. This technique is not so appropriate if there are more than two equations or if the equations are non-linear. We will briefly describe the phase plane techniques for dealing with two coupled differential equations for $x = x(t)$, $y = y(t)$ i.e.

$$\frac{dx}{dt} = F(x, y), \qquad \frac{dy}{dt} = G(x, y) \qquad (7.1)$$

where F and G are continuous functions of x and y and have continuous partial derivatives.

For a given set of initial conditions, $x(0) = x_0$, $y(0) = y_0$, (7.1) will have a unique solution $x = \phi(t)$, $y = \psi(t)$ for $t \geq 0$; and it is helpful to represent it as a curve in the $x - y$ plane, called the *phase plane* with t as parameter. The solution curves are referred to as *trajectories*.

As a simple example, consider the equation of a particle of mass m moving on along a horizontal table, attached by a spring to a fixed point, as illustrated in Fig. 5.1. With no resistance and external force, the governing equation, from equation (5.9), is

$$\frac{d^2x}{dt^2} = -\omega^2 x. \qquad (7.2)$$

Defining $y = dx/dt$, we have the system of differential equations

$$dx/dt = y, \quad dy/dt = -\omega^2 x. \qquad (7.3)$$

We have already found the solution, equation (5.11)

$$x = \alpha \cos(\omega t + \beta)$$

and from (7.3) we have

$$y = -\alpha\omega \sin(\omega t + \beta),$$

These equations define the trajectories

$$x^2 + (y/\omega)^2 = \alpha^2, \tag{7.4}$$

which are ellipses, in the $x - y$ plane, as shown in Fig. 7.1.

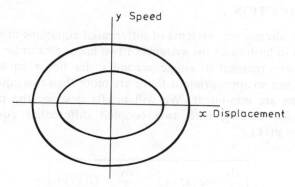

Fig. 7.1—Trajectories in x–y plane

We can obtain the trajectories directly from (7.3) by writing

$$\frac{dy}{dx} = \frac{dy/dt}{dx/dt} = -\frac{\omega^2 x}{y},$$

and we can separate variables to give

$$\int x \, dx + \int \frac{y}{\omega^2} \, dy = 0;$$

Integrating

$$\tfrac{1}{2}x^2 + \tfrac{1}{2}(y/\omega)^2 = k \text{ constant},$$

which gives (7.4) with $\alpha^2 = 2K$.

Returning to the general problem, equations (7.1), we see that

$$\frac{dy}{dx} = \frac{dy/dt}{dx/dt} = \frac{G(x, y)}{F(x, y)}, \quad \text{provided } F \neq 0. \tag{7.5}$$

Points (x_0, y_0) at which both F and G vanish i.e. $F(x_0, y_0) = G(x_0, y_0) = 0$ are called *equilibrium* or *critical* points. These points have special significance, for if $x = x_0$, $y = y_0$, from (7.1), $dx/dt = 0$, and so the trajectory remains at (x_0, y_0). So $x = x_0$, $y = y_0$ is a constant solution.

Determination of the equilibrium points and the behaviour of trajectories near them gives a great deal of information about the solutions, as we shall see. We state a number of important results, which can be proved;

(i)   through any point in the phase plane, there is at most one trajectory;

(ii) a trajectory starting at a point which is not an equilibrium point cannot reach an equilibrium point in finite time; so that if a trajectory approaches an equilibrium point (x_0, y_0), then necessarily $t \to \infty$;

(iii) a trajectory cannot cross itself, unless it is a closed curve, in which case the trajectory corresponds to a periodic solution of (7.1).

In order to investigate the behaviour near an equilibrium point we first assume that the equilibrium point is $x_0 = 0$, $y_0 = 0$ (this involves no loss of generality, since if $x_0 \neq 0$, $y_0 \neq 0$, we make the substitution $x = x_0 + u$, $y = y_0 + v$). Then for x and y small

$$\frac{dx}{dt} = F(x, y) = F(0, 0) + x \frac{\partial F}{\partial x}(0, 0) + y \frac{\partial F}{\partial y}(0, 0) + \text{higher order terms}$$

i.e.
$$\frac{dx}{dt} \approx ax + by \tag{7.6}$$

since $F(0, 0) = 0$. Similarly

$$\frac{dy}{dt} \approx cx + dy. \tag{7.7}$$

We can solve (7.6) and (7.7) by assuming $x = Ae^{rt}$, $y = Be^{rt}$, so that (7.6) gives

$$(r - a)A = bB$$

and (7.7) gives

$$(r - d)B = cA.$$

Equating the ratio A/B gives the quadratic equation

$$\boxed{r^2 - (a + d)r + ad - bc = 0} \quad . \tag{7.8}$$

Assuming $(ad - bc) \neq 0$, we have two non-zero values for r, say r_1 and r_2, and the general solution

$$x = A_1 e^{r_1 t} + A_2 e^{r_2 t}, y = B_1 e^{r_1 t} + B_2 e^{r_2 t}$$

where only two of the arbitrary constants, A_1, A_2, B_1 and B_2 are arbitrary, for $A_1/B_1 = b/(r_1 - a)$, $A_2/B_2 = b/(r_2 - a)$.

The behaviour depends on the values of r_1 and r_2. The table and diagrams below illustrate the behaviour for the possible cases.

As can be seen from the sketches, we define the equilibrium point $(a, 0)$ as

(i)   *stable* if x and y remain bounded as $t \to \infty$;
(ii) *Asymptotically stable* if $x, y \to 0$ as $t \to \infty$;
(iii) *Unstable* otherwise

Example
   Investigate the equilibrium points of the system of differential equations

$$\frac{dx}{dt} = x - x^2 - xy, \qquad \frac{dy}{dt} = \tfrac{1}{2}y - \tfrac{1}{4}y^2 - \tfrac{3}{4}xy. \tag{7.9}$$

Solution
   Equilibrium points are given by solutions of

$$x - x^2 - xy = 0 \text{ i.e. } x(1 - x - y) = 0$$

$$1/2y - 1/4y^2 - 3/4xy = 0 \text{ i.e. } y(1/2 - 1/4y - 3/4x) = 0$$

This gives four points, namely $(0, 0)$, $(1, 0)$, $(0, 2)$, $(1/2, 1/2)$.

$(0, 0)$ For small x and y, we have approximately

$$\frac{dx}{dt} = x, \qquad \frac{dy}{dt} = \tfrac{1}{2}y.$$

Roots	Name of Type of Equilibrium Point	Sketch	Stability
$r_1 < r_2 < 0$	Improper Node		asymptotically stable
$r_1 > r_2 > 0$			unstable
$r_2 < 0 < r_1$	Saddle Point		unstable
$r_1 = r_2 < 0$ $r_1 = r_2 > 0$	Node		asymptotically stable unstable
$r_1, r_2 = \lambda \pm i\mu$ $\lambda > 0$ $\lambda < 0$	Spiral Point		unstable asymptotically stable
$r_1 = i\mu, r_2 = -i\mu$	Centre		stable

real roots

complex roots

So $a = 1, b = 0, c = 0, d = 1/2$, and from (7.8),

$$r^2 = (3/2)r + (1/2) = 0 \text{ i.e. } r_1 = 1, r_2 = 1/2,$$

and we have an unstable node.

$(1, 0)$ we must first substitute $x = 1 + u, y = 0 + v$ giving

$$\frac{du}{dt} = (1 + u) - (1 + u)^2 - (1 + u)v$$

i.e. $\dfrac{du}{dt} = -u - v$,   neglecting u^2 and v^2 terms.

Similarly

$$\frac{dv}{dt} = -\frac{v}{4}, \quad \text{so that} \quad a = -1, b = -1, c = 0, d = -1/4$$

and (7.8) gives

$$r^2 + (5/4)r + (1/4) = 0 \text{ i.e. } r_1 = -1, r_2 = -1/4$$

and we have an asymptotically stable node.

Similarly $(0, 2)$ is an asymptotically stable node, and $(1/2, 1/2)$ is an

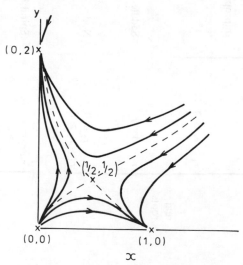

Fig. 7.2—Trajectories for equation (7.9)

unstable saddle point. A complete sketch of the trajectories is shown in Fig. 7.2.

7.2 INTERACTING SPECIES

Species can interact in many different ways. Before developing models to describe the interaction, we first look at one specific observation. The data given below gives the percentage of total catch of selachians (sharks, skates etc.), which are not very desirable fish for food, found in Mediterrean ports during 1914–1923.

Year	1914	1915	1916	1917	1918	1919	1920	1921	1922	1923
%	11.9	21.4	22.1	21.2	36.4	27.3	16.0	15.9	14.8	10.7

At first sight, it is rather surprising to see the large increase in the percentage of selachians during the war years. Now selachians are predators whilst food fish are their prey. The level of fishing during the war was much reduced, resulting in more prey available for the predators. The selachians therefore thrived, but what happened to the food fish is not so clear.

In mathematical terms we first define the prey and predator populations as x and y respectively. Now the governing differential equations for two-species interaction can be written as

$$\frac{dx}{dt} = f(x, y)$$

$$\frac{dy}{dt} = g(x, y).$$

For our example, we assume that in the absence of predators, prey will grow unlimited, according to $dx/dt = \alpha x$, whilst in the absence of prey, the predators will die out according to $dy/dt = -\delta y$. We model the interaction term by xy, positive for the predator, negative for the prey, resulting in the model

$$\boxed{\frac{dx}{dt} = \alpha x - \beta xy} \qquad (7.10)$$

$$\boxed{\frac{dy}{dt} = -\gamma y + \delta xy} \qquad (7.11)$$

where α, β, γ and δ are positive constants.

There are a number of ways we can go about solving these equations. For example we can write

$$\frac{dy}{dx} = \frac{dy}{dt} \Big/ \frac{dx}{dt} = \frac{(-\gamma + \delta x)y}{(\alpha - \beta y)x},$$ (7.12)

which is a first order variable separable differential equation; and we can solve to give

$$\int \frac{\alpha - \beta y}{y} \, dy = \int \frac{(-\gamma + \delta x)}{x} \, dx$$

i.e.

$$\alpha \ln y - \beta y = -\gamma \ln x + \delta x + K,$$

where K is the constant of integration. We can rewrite the solution as

$$\boxed{\frac{y^{\alpha} x^{\gamma}}{e^{\beta y + \delta x}} = K'},$$ (7.13)

where K' is a constant. This equation defines the $x - y$ solution trajectories, but it is not clear what they look like as y (or x) cannot be expressed as a function of x (or y).

So we return to (7.10) and (7.11) and use the phase plane techniques discussed in 7.1. We first note that there are *two* critical points

$$(0, 0) \quad \text{and} \quad (\gamma/\delta, \, \alpha/\beta)$$ (7.14)

Near the point $(0, 0)$, we can approximate (7.10) and (7.11) by

$$\frac{dx}{dt} \approx \alpha x, \qquad \frac{dy}{dt} \approx -\gamma y$$ (7.15)

and in the notation of (7.6) and (7.7) we have

$$a = \alpha, \, b = 0, \, c = 0, \, d = -\gamma.$$

Now (7.8) gives

$$r^2 - \alpha r - \alpha \gamma = 0$$

i.e.

$$r = \tfrac{1}{2}\alpha \pm \tfrac{1}{2}(\alpha^2 + 4\alpha\gamma)^{1/2}$$

This is the case $r_2 < 0 < r_1$ and from the table in 7.1 we have a 'saddle point' at $(0, 0)$.

For points near $(\gamma/\delta, \alpha/\beta)$, we put

$$x = \gamma/\delta + u, y = \alpha/\beta + v \qquad (7.16)$$

where u and v are small, in (7.10) and (7.11). For (7.10), we obtain

$$\frac{d}{dt}(\gamma/\delta + u) = \alpha(\gamma/\delta + u) - \beta(\gamma/\delta + u)(\alpha/\beta + v)$$

i.e.

$$\frac{du}{dt} \approx - (\beta\gamma/\delta)v, \qquad (7.17)$$

neglecting the 'uv' term. Similarly (7.11) gives

$$\frac{dv}{dt} = (\delta\alpha/\beta)u, \qquad (7.18)$$

and again using the notation of (7.6) and (7.7), $x = u$, $y = v$ and $a = 0$,
$$b = -\beta\gamma/\delta, c = \delta\alpha/\beta, d = 0;$$

so that (7.8) gives

$$r^2 + \alpha\gamma = 0.$$

Thus $r = \pm i(\alpha\gamma)^{1/2}$ and from the table in 7.1 we have a 'centre'.

We can now begin to sketch the trajectories. Fig. 7.3a illustrates the behaviour near the critical points, whereas Figure 7.3b shows a full sketch, deduced from Fig. 7.3a. Note that we can deduce the direction of motion by looking at sign of dx/dt and dy/dt. These trajectories can of course be found by numerical evaluation of the solution (7.13).

The first property of the trajectories to notice is that they are *closed*, so that the solutions are periodic with time and predict that neither predator or prey ever becomes extinct. Secondly, each trajectory can be divided into four distinct regions (see Fig. 7.3b):

(i)   In this region, the number of selachians decrease because of the lack of food fish, whereas the food fish population can increase due to lack of predators.

(ii) Here the food fish population has increased so much that the selachians population can also increase.

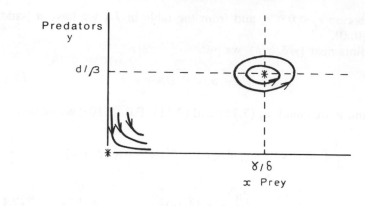

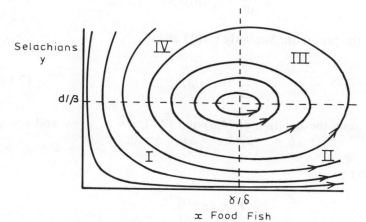

Fig. 7.3—Closed Trajectories for Predator–Prey System

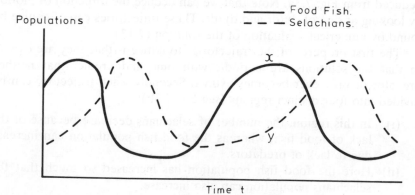

Fig. 7.4—Time History of Populations

(iii) Now the predators have increased so much that the food fish population is in decline.

(iv) Due to lack of food fish, both predator and prey are in decline.

The time histories of each population are sketched in Fig. 7.4 (these solutions have to be found numerically)

In order to assess the effect of harvesting, we first evaluate the *average* value of the prey and predator over a whole cycle. These are defined as

$$\bar{x} = \frac{1}{T}\int_0^T x(t)\,dt, \qquad \bar{y} = \frac{1}{T}\int_0^T y(t)\,dt, \qquad (7.19)$$

where T is the period of the cycle. Now, from (7.10),

$$\frac{1}{x}\frac{dx}{dt} = \alpha - \beta y,$$

and integrating from $t = 0$ to $t = T$,

$$\int_0^T \frac{1}{x}\frac{dx}{dt}\,dt = \int_0^T (\alpha - \beta y)\,dt$$

i.e.

$$\int_0^T \frac{1}{x}\,dx = \alpha T - \beta \int_0^T y(t)\,dt,$$

$$\ln[x(T)/x(0)] = \alpha T - \beta \int_0^T y(t)\,dt.$$

But $x(T) = x(0)$, since it is a complete cycle; so that we obtain

$$0 = \alpha T - \beta \int_0^T y(t)\,dt$$

Hence, from (7.19),

$$\boxed{\bar{y} = \alpha/\beta}\ , \qquad (7.20)$$

and similarly, using (7.11),

$$\boxed{\bar{x} = \gamma/\delta}\ . \qquad (7.21)$$

So the *average* value of the predator and prey are in fact their *equilibrium* values.

Now fishing will decrease the food fish population at a rate $Ex(t)$, and decrease the selachian population at a rate $Ey(t)$, where E is the fishing effort (e.g. number of boats, nets, pots, etc.). The modified system of differential equations becomes

$$\frac{dx}{dt} = \alpha x - \beta xy - Ex = (\alpha - E)x - \beta xy \qquad (7.22)$$

$$\frac{dy}{dt} = -\partial y + \delta xy - Ey = -(\gamma + E)y + \delta xy. \qquad (7.23)$$

These equations are identical to (7.10) and (7.11) with α replaced by $(\alpha - E)$ and γ by $(\gamma + E)$. The *new* average values are, from (7.20) and (7.21), given by

$$\bar{y} = (\alpha - E)/\beta, \bar{x} = (\gamma + E)/\delta. \qquad (7.24)$$

This implies that a moderate amount of fishing $(E < \alpha)$ actually *increases* the average food fish population and *decreases* the average the number of selachians, the opposite of what might be expected. Conversely, a reduced level of fishing *increases* on average, the number of selachians and *decreases* the food fish population. This remarkable result (called Volterra's principle) explains the data at the beginning of this section.

A similiar result is obtained when this principle is applied to insecticide treatments, which destroy both insect predators and insect prey.

7.3 COMPETING SPECIES: THE STRUGGLE FOR EXISTENCE

In this section we consider not a predator–prey relationship but a two-species ecosystem in which both species compete for the same limited food supply. We start by considering what happens if only one of the species is present; in this case we assume the logistic model

$$\frac{dx}{dt} = ax - bx^2 \quad \text{if} \quad y = 0$$

where x and y are the species populations. Similarly

$$\frac{dy}{dt} = cx - dx^2 \quad \text{if} \quad x = 0.$$

The effect of competition is modelled by assuming the growth rate is reduced by a factor proportional to the other species population. Thus the governing equations for two competing species are

$$\frac{dx}{dt} = x(a - bx - my) \qquad (7.25)$$

$$\frac{dy}{dt} = y(c - dy - nx) \qquad (7.26)$$

where a, b, c, d, m and n are positive constants.

This system of coupled differential equations does not have an analytical solution, so we first find the critical points where $dx/dt = dy/dt = 0$. This gives the three points

$$(0, 0), \ (0, c/d), \ (a/b, 0); \qquad (7.27)$$

and also the solution of

$$bx + my = a$$

$$nx + dy = c$$

which, provided $bd - mn \neq 0$, is given by

$$\left(\frac{ad - bm}{bd - mn}, \frac{cb - an}{bd - mn} \right). \qquad (7.28)$$

This critical point could be located in any of the four quadrants of the phase plane depending on the values of the parameters.

We will just continue the analysis in the case of two competing species which are virtually identical; that is, we take

$$a = c, b = d.$$

We also assume that one of the two species is more suited for competition; for example we take $n > m$ which means that x is stronger.

The equations are now

$$\frac{dx}{dt} = x(a - bx - my)$$ (7.29)

$$\frac{dy}{dt} = y(a - by - nx)$$ (7.30)

and we have critical points

$$(0, 0), (0, a/b), (a/b, 0), (b(m - b)/(mn - b^2), a(n - b)/(mn - b^2))$$ (7.31)

The fourth critical point is in the positive xy quadrant of the phase plane. The critical points are illustrated in Fig. 7.5. Also we note that

$$\frac{dy}{dx} = \frac{y(a - by - nx)}{x(a - bx - my)}$$ (7.32)

and so

(i) $\dfrac{dy}{dx} = 0$ on $y = 0$ and $by + nx = a$

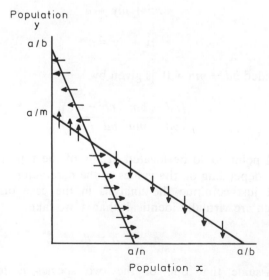

Fig. 7.5—Lines on which $\dfrac{dy}{dx} = 0$ or ∞

(ii) $\dfrac{dy}{dx} = \infty$ on $x = 0$ and $bx + my = a$.

The directions of the trajectories are sketched in Fig. 7.5.

We also note that for $x > 0$,

$$\frac{dx}{dt} \begin{cases} > 0 & \text{if} \quad a > bx + my \\ < 0 & \text{if} \quad a < bx + my \end{cases}$$

and so we can put arrows on the trajectories as shown.

We can also use the analysis of the critical points (from 7.1) to obtain sketches of the trajectories near the critical points, and so we are able to obtain a complete sketch of the trajectories as shown in Fig. 7.6.

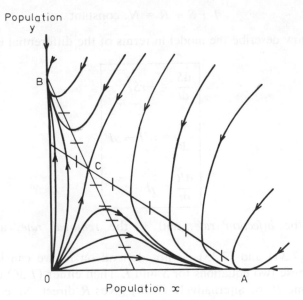

Fig. 7.6—Sketch of Trajectories

The equilibrium point C is unstable, so that coexistence is not possible. On the other hand both equilibrium points A and B in which one species is *extinct* are stable. So we conclude that in time one or other of the populations will become extinct. This is an example of the *principle of competitive exclusion*. Only one species can in the long run survive; and although one species, x, is stronger, the trajectories in Fig. 7.6 show that it is still possible for it to become extinct. It all depends on the initial conditions.

7.4 EPIDEMICS

In this section we consider the problem of an infectious disease which is introduced into a closed population. We want for example to estimate how many of the population will catch the disease?

We first assume that any individual who has recovered from the disease then has permanent immunity, and that the disease has a negligibly short incubation period. We divide the population into three classes,

(i) *infectives*, say I, who can transmit the disease;
(ii) *susceptibles*, say S, who are capable of catching the disease;
(iii) *removals*, say R, who have had the disease, and are dead or recovered and immune or isolated.

If N is the total population size, we must have

$$I + S + R = N, \text{ constant.} \tag{7.33}$$

We can now describe the model in terms of the differential equations

$$\frac{dS}{dt} = -rSI \tag{7.34}$$

$$\frac{dI}{dt} = rSI - \gamma I \tag{7.35}$$

$$\frac{dR}{dt} = \gamma I \tag{7.36}$$

Here r, the *infection rate*, and γ, the *removal rate*, are positive constants.

Since (7.34) and (7.35) do not depend on R, we can initially just consider these two equations for S and I. Then either (7.36) can be used to determine R, or alternatively (7.33) gives R direct. Note that (7.33) is a consequence of adding (7.34), (7.35) and (7.36), which gives

$$\frac{d}{dt}(S + I + R) = 0,$$

and integrating.

We can combine (7.34) and (7.35) to give

$$\frac{dI}{dS} = \frac{rSI - \gamma I}{-rSI} = -1 + \rho/S \tag{7.37}$$

where $\rho = \gamma/r$, and integrating

$$I = -S + \rho \log S + K.$$

If at $t = 0$, $I = I_0$ and $S = S_0 = N = I$. $(R_0 = 0)$, then

$$K = N - \rho \log S_0,$$

giving

$$\boxed{I = N - S + \rho \log(S/S_0)} \qquad (7.38)$$

Now, from (7.37), dI/ds is positive (negative) for $\rho/s > 1$ $(\rho/s < 1)$ i.e. I is an increasing (decreasing) function for $S < \rho$ $(S > \rho)$. Also $I \to -\infty$ as $S \to 0$ and $I = I_0$ (> 0) when $S = S_0$. Hence there exists at least one point, say S_∞, when $I = 0$. In fact S_∞ is a unique point and from (7.34) and (7.35) we can see that $S = S_\infty$, $I = 0$ is an equilibrium point. Typical trajectories in the $S - I$ plane are sketched in Fig. 7.7.

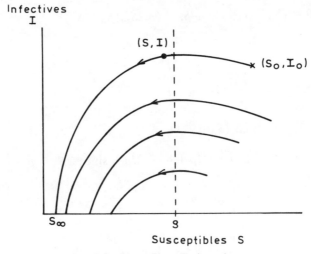

Fig. 7.7—Phase Plane Trajectories

We see that as t increases, the point (S, I) moves along the trajectory with S decreasing, and I decreases monotonically if $S_0 < \rho$. On the other hand, if $S_0 > \rho$, then initially I increases reaching a peak when $S = \rho$, and then decreasing to zero. Clearly the parameter ρ is a vital parameter. It is called the *threshold value*, since an epidemic occurs only if the initial number of susceptibles, S_0, is greater than the threshold value.

Assuming $S_0 > \rho$, we can deduce the Threshold Theorem of Epidemiology, which states that if $S_0 - \rho$ is small compared to ρ, then the number of individuals who ultimately contract the disease is approximately $2(S_0 - \rho)$.

To prove this result, let $S_0 = \rho + v$, where $v << \rho$ (i.e. v is very much smaller than ρ). We also assume that I_0, the initial number of infectives, is small. Now, from (7.38), as $t \to \infty$,

$$0 = N - S_\infty + \rho \log(S_\infty/S_0)$$

and $N \simeq S_0$. Thus

$$0 = S_0 - S_\infty + \rho \log\left[\frac{S_0 - (S_0 - S_\infty)}{S_0}\right]$$

$$= S_0 - S_\infty + \rho \log[1 - (1 - S_\infty/S_0)]$$

$$= S_0 - S_\infty + \rho\{(1 - S_\infty/S_0) - (1/2)(1 - S_\infty/S_0)^2 \ldots\}$$

i.e. $0 = (S_0 - S_\infty)[1 - \rho/S_0 - (\rho/2S_0^2)(S_0 - S_\infty)].$

Thus

$$S_0 - S_\infty = \frac{2S_0^2}{\rho}(1 - \rho/S_0)$$

$$= 2S_0(S_0/\rho - 1)$$

$$= 2(\rho + v)(\rho + v - \rho)/\rho, \quad \text{since} \quad S_0 = \rho + v$$

$$= 2\frac{v}{\rho}(\rho + v)$$

$$\approx 2\frac{v}{\rho}\rho, \quad \text{since} \quad v << \rho$$

i.e. $\boxed{S_0 - S_\infty \approx 2v}$,   (7.39)

which proves the result. Fig. 7.8 indicates the situation.

Public health statistics usually record the number of new removals each day or week, so to compare the predicted results of this model with actual data we must determine dR/dt as a function of time, t. Now, from

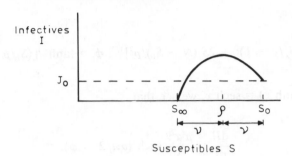

Fig. 7.8—The Threshold Theorem of Epidemiology

(7.36) and (7.33)

$$\frac{dR}{dt} = \gamma(N - R - S),\qquad (7.40)$$

whilst we can determine S as a function of R from

$$\frac{dS}{dR} = \frac{dS/dt}{dR/dt} = -\frac{rSI}{\gamma I} = -\frac{S}{\rho}.$$

Solving for S gives $S = S_0 e^{-R/\rho}$, and substituting in (7.40) gives

$$\frac{dR}{dt} = \nu(N - R - S_0 e^{-R/\rho}).\qquad (7.41)$$

Unfortunately this cannot be solved explicity, but assuming R/ρ is reasonably small,

$$e^{-R/\rho} \simeq 1 - R/\rho + (R/\rho)^2/2 + \ldots \ldots$$

so that

$$\frac{dR}{dt} \simeq \nu[N - R - S_0(1 - R/\rho + R^2/2\rho^2)]$$

i.e.

$$\frac{dR}{dt} = \nu[N - S_0 + (S_0/\rho - 1)R - S_0 R^2/2\rho^2].\qquad (7.42)$$

It can be verified that this equation has solution

$$R = \rho^2[S_0/\rho - 1 + \alpha \tanh(\alpha \nu t/2 - \phi)]/S_0\qquad (7.43)$$

where

$$\alpha = [(S_0/\rho - 1)^2 + 2S_0(N - S_0)/\rho^2]^{1/2}, \quad \phi = \tanh^{-1}[(S_0/\rho - 1)/\alpha]$$

Since $\dfrac{d}{dx}(\tanh x) = \operatorname{sech}^2 x$, we see that

$$\frac{dR}{dt} = \frac{\rho^2 \alpha^2 v}{2S_0} \operatorname{sech}^2(\alpha v t/2 - \phi). \qquad (7.44)$$

This defines a bell-shaped curve, known as the epidemic curve. Its maximum value occurs at $t = 2\phi/\alpha v$, and is illustrated in Fig. 7.9.

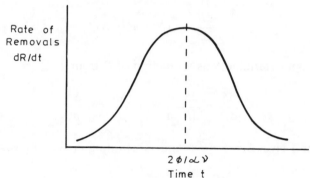

Fig. 7.9—Predicted Rate of Removals

Kermack and McKendrick used this model to describe a plague in Bombay. They took parameter values so that

$$\frac{dR}{dt} = 890 \operatorname{sech}^2(0.2t - 3.4), \qquad (7.45)$$

with t measured in weeks. We can compare dR/dt with the number of deaths per week, since almost all cases terminated fatally. Fig. 7.10 illustrates the predicted curve (7.45) with the actual data.

7.5 SPRING MASS SYSTEM

In Fig. 7.11 we illustrate a double spring mass system. We will be concerned with oscillations in a vertical plane, the upper end of the upper spring being fixed. If you have suitable equipment it would be beneficial to set up the system as illustrated, and giving small pushes to the masses, observe what happens to the motion of the masses.

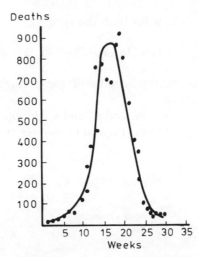

Fig. 7.10—Real data and Theoretical Curve

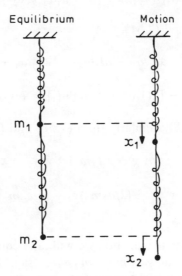

Fig. 7.11—Spring-mass System

We now construct a theoretical model based on Newton's Laws of Motion in order to predict the nature of the oscillations observed. In equilibrium, the forces on both m_1 and m_2 balance giving

$$T_1 = m_1 g + T_2 \tag{7.46}$$

$$T_2 = m_2 g \tag{7.47}$$

and assuming Hook's Law for both the springs,

$$T_1 = \lambda d_1/a_1, \quad T_2 = \lambda d_2/a_2 \qquad (7.48)$$

Here a_1, a_2 are the natural lengths of the springs, and d_1, d_2 are the extensions in equilibrium.

If the system is in motion, and x_1 and x_2 denote the displacements of the masses from equilibrium, then applying Newton's Laws of Motion, we have

$$m_1\ddot{x}_1 = m_1g + T_2 - T_1 \qquad (7.49)$$

$$m_2\ddot{x}_2 = m_2g - T_2 \qquad (7.50)$$

where

$$T_1 = \lambda(d_1 + x_1)/a_1, \quad T_2 = \lambda(d_2 + x_2 - x_1)/a_2. \qquad (7.51)$$

Hence, from (7.49) and (7.51),

$$m_1\ddot{x}_1 = (m_1g + \lambda d_2/a_2 - \lambda d_1/a_1) - \lambda x_1/a_1 + \lambda(x_2 - x_1)/a_2$$

i.e.
$$\ddot{x}_1 = -[\lambda(1/a_1 + 1/a_2)/m_1]x_1 + (\lambda/a_2m_1)x_2, \qquad (7.52)$$

using (7.46) and (7.48). Similarly from (7.50) and (7.51),

$$m_2\ddot{x}_2 = (m_2g - \lambda d_2/a_2) - \lambda(x_2 - x_1)/a_2$$

i.e.
$$\ddot{x}_2 = -(\lambda/a_2m_2)x_2 + (\lambda/a_2m_2)x_1, \qquad (7.53)$$

using (7.47) and (7.48).

In order to simplify the algebra, we will take equal springs, i.e. $a_1 = a_2 = a$, say, and equal masses $m_1 = m_2 = m$, say. Then, with $k = \lambda/am$, (7.52) and (7.53) reduce to

$$\ddot{x}_1 = -2kx_1 + kx_2 \qquad (7.54)$$

$$\ddot{x}_2 = -kx_2 + kx_1 \qquad (7.55)$$

(7.54) and (7.55) constitute a pair of coupled *second* order differential equations. There are a number of ways in which we can solve these. For example, one way is to define $x_3 = \dot{x}_1$, and $x_4 = \dot{x}_2$ and consider the set of four coupled first order differential equations

$$\dot{x}_1 = x_3$$

$$\dot{x}_2 = x_4 \tag{7.56}$$

$$\dot{x}_3 = -2kx_1 + kx_2$$

$$\dot{x}_4 = kx_1 - kx_2$$

A more direct method is to assume an oscillating type of solution and then determine its characteristics. For example, if

$$x_1 = A_1 \cos \omega t, \ x_2 = A_2 \cos \omega t, \tag{7.57}$$

(7.54) and (7.55) reduce to

$$(\omega^2 - 2k)A_1 + kA_2 = 0$$

$$kA_1 + (\omega^2 - k)A_2 = 0$$

For non-trivial solutions for A_1, A_2 we require

$$\begin{vmatrix} \omega^2 - 2k & k \\ k & \omega^2 - k \end{vmatrix} = 0$$

i.e.

$$\boxed{\omega^4 - 3k\omega^2 + k^2 = 0} \tag{7.58}$$

Solving

$$\omega_1^2 = (3 + \sqrt{5})k/2, \ \omega_2^2 = (3 - \sqrt{5})k/2,$$

giving two positive real values for ω. For $\omega = \omega_1$, we find that

$$\frac{A_1}{A_2} = -\frac{(1 + \sqrt{5})}{2}$$

and for $\omega = \omega_2$,

$$\frac{A_1}{A_2} = \frac{(\sqrt{5} - 1)}{2}.$$

Thus we have two distinct solutions, namely

$$\left.\begin{aligned} x_1 &= -\frac{(1 + \sqrt{5})}{2} k_1 \cos \omega_1 t \\[2mm] x_2 &= k_1 \cos \dot{\omega}_1 t \end{aligned}\right\} \tag{7.59}$$

and

$$\left.\begin{aligned} x_1 &= \frac{(\sqrt{5} - 1)}{2} k_2 \cos \omega_2 t \\[2mm] x_2 &= k_2 \cos \omega_2 t \end{aligned}\right\} \tag{7.60}$$

These two solutions correspond to the *normal modes of the system*. In the first, $\omega = \ddot{\omega}_1$, both masses perform S.H.M. with the same period, but, since $x_1/x_2 = -(1 + \sqrt{5})/2$, in *opposing* directions with the amplitude of m_1 larger than m_2's. In the second mode, $\omega = \omega_2$, again the masses perform S.H.M. but this time in the *same* direction with m_1's amplitude smaller than m_2's. Both modes are illustrated in Fig. 7.12.

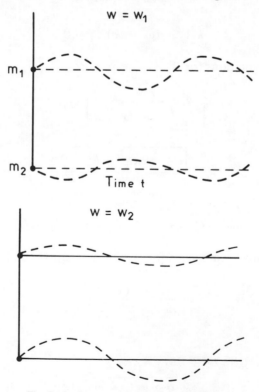

Fig. 7.12—Normal Modes of Oscillations

We have deduced the general characteristics of the motion. The system will tend to oscillate in one of the two normal modes, and experiments will agree with this. On the other hand, to predict the actual time history of the system, we must consider specific initial conditions

$$x_1 = 0, \dot{x}_1 = 0, x_2 = b, \dot{x}_2 = 0 \qquad (7.61)$$

at $t = 0$, and solve the problem completely.

A general solution of (7.54) and (7.55) will also include sine terms and will be of the form

$$x_1 = -\frac{(1+\sqrt{5})}{2}(k_1 \cos \omega_1 t + k_3 \sin \omega_1 t) + \frac{(\sqrt{5}-1)}{2}(k_2 \cos \omega_2 t + k_4 \sin \omega_2 t)$$

$$x_2 = k_1 \cos \omega_1 t + k_3 \sin \omega_1 t + k_2 \cos \omega_2 t + k_4 \sin \omega_2 t.$$

Applying conditions (7.61) gives

$$0 = -\frac{(1+\sqrt{5})}{2}k_1 + \frac{(\sqrt{5}-1)}{2}k_2$$

$$0 = -\frac{(1+\sqrt{5})}{2}\omega_1 k_3 + \frac{(\sqrt{5}-1)}{2}\omega_2 k_4$$

$$b = k_1 + k_2$$

$$0 = \omega_1 k_3 + \omega_2 k_4$$

and, solving,

$$k_1 = (\sqrt{5}-1)b/2\sqrt{5}, k_2 = (1+\sqrt{5})b/2\sqrt{5}, k_3 = k_4 = 0.$$

This gives the complete solution

$$x_1 = b[-\cos \omega_1 t + \cos \omega_2 t]/\sqrt{5} \qquad (7.62)$$

$$x_2 = b[(\sqrt{5}-1)\cos \dot{\omega}_1 t + (1+\sqrt{5})\cos \omega_2 t]/2\sqrt{5} \qquad (7.63)$$

7.6 THE DYNAMICS OF ARMS RACES

In our final section we describe a mathematical model for the build-up of conflict between two nations. Each nation is ready to defend itself against the other and each considers the possibilities of attack are real. To formulate a mathematical model, we first define $x(t)$ and $y(t)$ as the 'war potential' of the two nations, say A and B. We can measure this potential in terms of, for example, the level of armaments of each country.

We now assume that the rate of increase of x depends linearly on the level y i.e. $dx/dt = ky$; we also suppose that the cost of increasing and maintaining armaments has a restraining effect so that we take $dx/dt = ky - \alpha x$; and finally we add a constant term which represents the underlying grievances felt by nation A towards nation B. This leads to the differential equation

$$\boxed{\frac{dx}{dt} = ky - \alpha x + g} \; ; \qquad (7.64)$$

similarly

$$\boxed{\frac{dy}{dt} = lx - \beta y + h} \; . \qquad (7.65)$$

We may deduce a number of important results from our model:

(i) Suppose there are no underlying grievances (i.e. $g = h = 0$); then $x = y = 0$ is an equilibrium point of (7.64) and (7.65). This ideal condition is permanent *peace* with disarmament and satisfaction. It has, for example, existed since 1817 on the border between Canada and the United States.

(ii) Mutual disarmament without satisfaction is not permanent since if $x = y = 0$ at some time, then immediately after

$$\frac{dx}{dt} = g, \qquad \frac{dy}{dt} = h,$$

and so x and y will not remain at zero if g and h are positive.

(iii) An armaments race occurs when the 'defence' terms, ky and lx, predominate in (7.64) and (7.65). In this case

$$\frac{dx}{dt} = ky, \qquad \frac{dy}{dt} = lx. \qquad (7.66)$$

Now

$$\frac{d^2x}{dt^2} = k\,\frac{dy}{dt} = klx,$$

and so x takes the form

$$x = Ae^{\sqrt{klt}} + Be^{-\sqrt{klt}}. \tag{7.67}$$

Also, from (7.66),

$$y = \frac{1}{k}\frac{dx}{dt} = \sqrt{\frac{l}{k}}\,(Ae^{\sqrt{klt}} - Be^{-\sqrt{klt}}). \tag{7.68}$$

These solutions show that x and y both approach infinity (if A is positive), which can be interpreted as war!

Returning to the general problem, described by (7.64) and (7.65), we see that there is a single equilibrium point

$$x_0 = \frac{kh + g\beta}{\alpha\beta - kl}, \qquad y_0 = \frac{\alpha h + gl}{\alpha\beta - kl}, \tag{7.69}$$

provided $(\alpha\beta - kl) \neq 0$. To determine its stability, put

$$x = x_0 + u, \, y = y_0 + v$$

where u and v are small changes from the equilibrium position, in (7.64) and (7.65). This gives

$$\frac{du}{dt} = -\alpha u + kv \tag{7.71}$$

$$\frac{dv}{dt} = lu - \beta v. \tag{7.72}$$

Equation (7.8) becomes

$$r^2 + (\alpha + \beta)r + \alpha\beta - lk = 0$$

giving

$$r = -(\alpha + \beta)/2 \pm ((\alpha + \beta)^2 - 4(\alpha\beta - lk))^{1/2}/2$$
$$= -(\alpha + \beta)/2 \pm ((\alpha - \beta)^2 + 4lk)^{1/2}/2.$$

Now both roots are real, and if $(\alpha\beta - lk) > 0$, both roots are negative so we have stability (see Table in section 7.1); but if $\alpha\beta - lk < 0$, one root will be positive (and one negative) giving an unstable equilibrium point.

We now make some estimates of the parameters. If $y = g = 0$, then $dx/dt = -\alpha x$, and so $x(t) = x_0 e^{-\alpha t}$. In this case

$$x(\alpha^{-1}) = x_0/e$$

and we take α^{-1} to be the lifetime of the nation's parliament. Estimations of k and l vary, but for an industrial nation $k = 0.3$ has been used.

For the European arms race of 1909–1914, France was allied with Russia and Germany with Austria–Hungary whilst neither Britain or Italy was in any definite alliance with either group. So let A represent Germany with Austria and Hungary, and B France and Russia. We take

$$k = l = 0.9, \ \alpha = \beta = 0.2 \tag{7.74}$$

(so $\alpha^{-1} = 5$ years). Then the equilibrium point is

$$\left(\frac{kh + \alpha g}{\alpha^2 - k^2}, \ \frac{\alpha h + kg}{\alpha^2 - k^2} \right)$$

and

$$\alpha\beta - kl = \alpha^2 - k^2 = (0.2)^2 - (0.9)^2 = -0.77. \tag{7.75}$$

This means that the equilibrium point is *unstable*, in agreement with the historical fact of war between these alliances.

We can provide a more positive verification of this model by adding (7.64) and (7.65) to give (when $k = l$, $\alpha = \beta$),

$$\frac{d}{dt}(x + y) = (k - \alpha)(x - y) + g + h. \tag{7.76}$$

We interpret x as $x = u - u_0$ where u is the defence budget and u_0 is the amount of goods exported between the two alliances (which tends to produce cooperation). Similarly $y = v - v_0$, so that (7.76) becomes

$$\frac{d}{dt}(u + v) = (k - \alpha)\left\{ u + v - \left[u_0 + v_0 - \frac{(g + h)}{(k - \alpha)} - \frac{1}{(k - \alpha)} \frac{d}{dt}(u_0 + v_0) \right] \right\}.$$

The defence budgets for the two alliances are given in the table below, and Fig. 7.13 plots the annual increments in $(u + v)$, $\Delta(u + v)$, against the average of $u + v$.

Defence Budgets (£ millions)

	1909	1910	1911	1912	1913
France	48.6	50.9	57.1	63.2	74.7
Russia	66.7	68.5	70.7	81.8	92.0
Germany	63.1	62.0	62.5	68.2	95.4
Austria–Hungary	20.8	23.4	24.6	25.5	26.9
Total $(u + v)$	199.2	204.8	214.9	238.7	289.0
$\Delta(u + v)$		5.6	10.1	23.8	50.3
Average value of $(u + v)$		202.0	209.8	226.8	263.8

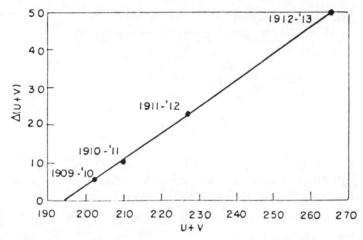

Fig. 7.13—Annual Increments in total Defence Budget. against total Defence Budgets

It is remarkable how close the data points are to the straight line

$$\Delta(u + v) = (0.73)(u + v - 194) \qquad (7.77)$$

Also the value $(k - \alpha) = 0.73$ is in good agreement with the value estimated earlier, (7.74).

Although there does appear good agreement between the theoretical model and actual data, it would not be reasonable to use such models to predict when a war will occur. On the other hand, it does present a clear picture as to what can happen if nations do not take steps to disarm and cooperate.

EXERCISES

1. **Interacting Species**
   (i) Predator–Prey

   A simple model for a predator–prey interaction is given by

   $$\frac{dx}{dt} = x + y$$

   $$\frac{dy}{dt} = -x + y.$$

   If the initial populations are $x(0) = y(0) = 1000$, determine x and y at future times. When does the prey become extinct?.

   (ii) Species Cooperation

   A model for species cooperation is given by

   $$\frac{dx}{dt} = -2x + 4y$$

   $$\frac{dy}{dt} = x - 2y.$$

   For initial populations (i) $x = 100$, $y = 300$ (ii) $x = 300$, $y = 100$, determine the predicted future populations and illustrate them graphically. What happens as $t \to \infty$?

   (iii) Species Competition

   This model describes two species who in the absence of each other increase in a Malthusian manner, but who kill each other when competing for food. The simplest form to take is

   $$\frac{dx}{dt} = ax - bxy$$

   $$\frac{dy}{dt} = cy + dxy.$$

   Determine equilibrium points, and behaviour near them, and sketch the trajectories in the $x - y$ plane. What conclusions can you make from the model.

(iv) Canadian lynx and snowshoe hare

Facts about the Canadian lynx and snowshoe hare

The cycles of both hares and lynxes have their peaks and troughs in approximately the same year all over Canada—there is no evidence that migration of the animals from one region to another could cause these fluctuations.

The lynx is known to have a diet over 90% of which is the hare. When the lynxes cannot obtain hares they starve rather than turn to other species.

In years when the population of hares falls dramatically, observers have noted that many of the dead hares are diseased and that the fall in number occurs mainly in the summer months.

In years when the population of hares is high, hares are found in areas which are not their natural habitat, but when their number is low, they are found almost exclusively in their natural habitat. Even in areas where lynxes have been virtually eliminated by man, the population of hares goes through a cyclic fluctuation.

Develop a mathematical model to describe the situation above, and compare its theoretical preditions with the data given.

2. **Spring–Mass Systems**

(i) Use the model in section 7.5 to find the normal modes of oscillation for two unequal masses m_1 and m_2.

(ii) Extend the model of section 7.5 to the case of three equal springs and three equal masses.

(iii) A spring-mass system, illustrated in Fig. 7.14, lies in a smooth horizontal table, the ends A and D being fixed. The particles at B and C both have mass m, and the springs AB, BC, and CD each have natural length $a/2$ and modulus λ.

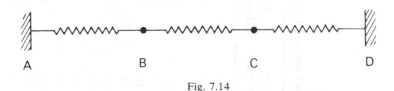

$$\begin{array}{cccc} A & B & C & D \end{array}$$

Fig. 7.14

If $AD = 3a$, show that for oscillations along the line of the springs, there are two possible frequencies, and describe each mode of oscillation.

The numbers of hare and lynx 1844–1934 (taken from the number of pelts sold to the Hudson Bay Trading Co., Canada)

Year	Hare (thousands)	Lynx (thousands)	Year	Hare (thousands)	Lynx (thousands)	Year	Hare (thousands)	Lynx (thousands)
1844	30	6	1875	105	26	1906	20	29
5	25	14	6	85	29	7	—	7
6	—	22	7	60	21	8	—	2
7	25	36	8	15	11	9	25	2
8	15	29	9	10	10	1910	50	4
9	30	7	1880	15	5	1	55	10
1850	55	2	1	10	3	2	75	14
1	80	1	2	10	5	3	70	19
2	80	1	3	40	16	4	55	—
3	90	1	4	50	42	5	30	8
4	70	5	5	135	64	6	20	9
5	80	13	6	135	63	7	15	2
6	95	16	7	90	32	8	15	1
7	75	25	8	30	15	9	20	1
8	30	14	9	20	7	1920	35	2

Year	Hare	Lynx		Year	Hare	Lynx		Year	Hare	Lynx
9	15	8		1890	50	3		1	60	4
1860	20	3		1	55	4		2	80	4
1	40	2		2	60	—		3	85	8
2	5	1		3	55	—		4	60	7
3	155	3		4	80	—		5	30	9
4	140	10		5	95	—		6	20	7
5	105	27		6	50	15		7	10	4
6	45	58		7	15	7		8	5	3
7	20	30		8	5	2		9	5	2
8	5	26		9	5	3		1930	10	3
9	5	9		1900	15	5		1	30	3
1870'	10	4		1	5	14		2	80	5
1	10	2		2	10	27		3	100	7
2	60	2		3	50	47		4	80	7
3	50	6		4	70	54				
4	50	10		5	—	—				

Note: — indicates no data available.
Numbers of hare are quoted to the nearest 5,000.
Numbers of lynx are quoted to the nearest 1,000.

3. **Arms Races**
Develop a 3-nation model to describe the current U.S.A./China/Russia situation. What conclusions can you make from your model?

4. **Protein Synthesis**
Protein synthesis can be modelled by the equations

$$\frac{dy}{dt} = \frac{c}{a + lz} - ky$$

$$\frac{dz}{dt} = ey - fz.$$

Here a, c, e, f, k and l are positive constants, and y is the RNA concentration and z the enzyme concentration.

The problem is to determine whether this model predicts sustained oscillations or decay for perturbations away from the equilibrium position.

References

[1] Andrews, J. G. and McLone, R. R., 1976, *Mathematical Modelling* (Butterworth).

[2] Aris, R., 1978, *Mathematical Modelling Techniques* (Pitman).

[3] Bender, E. A., 1978, *An Introduction to Mathematical Modelling* (Wiley).

[4] Bittinger, M. L., 1976, *Calculus, A. Modelling Approach* (Addison–Wesley).

[5] Braun, M., 1975, *Differential Equations and their Applications* (Springer Verlag).

[6] Burghes, D. N. and Wood, A. D., 1979, *Mathematical Models in the Social, Managerial and Life Sciences* (Ellis Horwood).

[7] Derrick, W. R. and Grossman, S., 1977, *Elementary Differential Equations with Applications* (Addison–Wesley).

[8] Haberman, R., 1977, *Mathematical Models* (Prentice-Hall).

[9] Holt, M. and Marjoram, D. T. E., 1973, *Mathematics in a Changing World* (Heinemann).

[10] Hull, J., Mapes, J. and Wheeler, B., 1976, *Model Building Techniques for Management* (Saxon House).

[11] Lancaster, P., 1976, *Mathematical Models of the Real World* (Prentice-Hall).

[12] Lighthill, J., 1978, *Newer Uses of Mathematics* (Penguin).

[13] McDonald, T., 1974, *Mathematical Methods for Social and Management Scientists* (Houghton Mifflin).

[14] Maki, D. P. and Thompson, M., 1973, *Mathematical Models and Applications* (Prentice-Hall).

[15] Mathematics Applicable Series (Discussion Unit, Starter Units and Continuation Units) (London, Heinemann).

[16] Olnick, M., 1978, *An Introduction to Mathematical Models in the Social and Life Sciences* (Addison–Wesley).

[17] Open University, 1977, Course TM281 'Modelling by Mathematics' (O.U. Press).

[18] Open University, 1978, Course MIOI, Block V. 'Mathematical Modelling' (O.U. Press).

[19] Roberts, C. E., 1979, *Ordinary Differential Equations* (Prentice-Hall).

[20] Bailey, N. T. J., 1976, *The Mathematical Theory of Infectious Diseases and Its Applications* (Hafner).

[21] Burghes, D. N. and Downs, A. M., 1975, *Modern Introduction to Classical Mechanics and Control* (Ellis Horwood).

[22] Burley, D., 1975, 'Mathematical Model for a Kidney Machine' *Mathematical Spectrum, 8* 69.

[23] Cope, F. W., 1976, 'Derivation of Weber Fechner Law and the Lowenstein Equation' *Bull. Math. Biol.* **38**, 111.

[24] Copemans, P., 1949, *Van Meegeren's Faked Vermeers and de Hooghs* (Meulenhoff).

[25] Gandolfo, G., 1971, *Mathematical Methods and Models in Economics and Dynamics* (North Holland).

[26] Grossman, S. and Turner, J., 1974, *Mathematics for the Biological Sciences* (Macmillan).

[27] Hutchinson, G. E., 1978, *An Introduction to Population Ecology* (Yale).

[28] Keisch, B., 1968, 'Dating Works of Art through their Natural Radioactivity: Improvements and Applications' *Science,* **160**, 413.

[29] Malthus, T. R., 1798, 'An Essay on the Principle of Population', (1926, reprinted Macmillan).

[30] Mansfield, E., 1961, 'Technical Change and the rate of imitation' Econometrics, *29*.

[31] Maynard-Smith, J., 1968, *Mathematical Ideas in Biology* (C.U.P.).

[32] Murdick, R., 1971, *Mathematical Models in Marketing* (Indext).

[33] Notari, R. E., 1975, *Biopharmaceutics and Pharmacokinetics* (Dekker).

[34] Pielou, E. C., 1969, *An Introduction to Mathematical Ecology* (Wiley).

[35] Rainey, R. H., 1967, 'National Displacement of Pollution from the Great Lakes' *Science,* **195**, 1242.

[36] Richardson, L. F., 1939, 'Generalised Foreign Politics' *Brit. J. Psychology*, monograph suppl. *23*.

Important Section References

Chapter One
1.1 [1]–[19]   1.2 [5]   1.3 [19]

Chapter Two
2.2 [9], [33]   2.3 [5], [14]   2.4 [17]   2.5 [4]   2.6 [22]

Chapter Three
3.2 [4], [23]   3.3 [21]   3.4 [18]   3.5 [27], [34]   3.6 [30]

Chapter Four
4.2 [32]   4.3 [5], [24], [28]   4.4 [7]   4.5 [17]   4.6 [25]   4.7 [35]

Chapter Five
5.2 [21]   5.3 [32]   5.4 [7]   5.5 [5]

Chapter Six
6.2 [21], [8]   6.3 [21], [7]   6.4 [7]

Chapter Seven
7.2 [5], [8], [56]   7.3 [5], [8], [27]   7.4 [19]   7.5 [21]   7.6 [35]

Index

Mathematics and its Applications

Series Editor: G. M. BELL, Professor of Mathematics, King's College London, University of London

Statistics, Operational Research and Computational Mathematics
Editor: B. W. CONOLLY, Emeritus Professor of Mathematics (Operational Research), Queen Mary College, University of London